Fluidization-dynamics

Fluidization-dynamics

The formulation and applications of a predictive theory for the fluidized state

L.G. Gibilaro

University of L'Aquila,
L'Aquila, Italy

OXFORD AUCKLAND BOSTON JOHANNESBURG MELBOURNE NEW DELHI

Butterworth-Heinemann
Linacre House, Jordan Hill, Oxford OX2 8DP
225 Wildwood Avenue, Woburn, MA 01801-2041
A division of Reed Educational and Professional Publishing Ltd

A member of the Reed Elsevier plc group

First published 2001
Transferred to digital printing 2004

British Library Cataloguing in Publication Data
A catalogue record for this book is available from the British Library

Library of Congress Cataloguing in Publication Data
A catalogue record for this book is available from the Library of Congress

ISBN 0 7506 5003 6

Typeset in India by Integra Software Services Pvt Ltd, Pondicherry, India 605005; www.integra-india.com

Contents

Preface

This book is intended for scientists and engineers who find themselves involved, for reasons ranging from pure academic interest to dire industrial necessity, in problems concerning the fluidized state. It has been written with two purposes in mind. The first is to present an analysis directed at the prediction of fluidized bed behaviour in systems for which empirical data is limited or unavailable. This represents a relevant goal; because alongside the advantages in the choice of a fluidized environment for achieving a processing objective there exist worrying uncertainties regarding the precise nature of the fluidization that will ensue; particles free of direct constraints on their positions and trajectories may well comport themselves in a manner that is highly disadvantageous to the purpose for which they are employed. Such occurrences are not unknown; disastrous mistakes have been made in the past, which inhibit the adoption of appropriate process solutions in the present.

The second objective is to provide a treatment of fluidization-dynamics that is readily accessible to the non-specialist. A stray encounter with the fluid-dynamic literature on the subject can be a disconcerting experience for the engineers seeking to improve their effectiveness in the practical application of fluidization technology. The linear approach adopted in this book, starting with the formulation of predictive expressions for the primary forces acting on a fluidized particle, is aimed at providing a clear route into the theory, and the incorporation of the force terms in the conservation equations for mass and momentum, and subsequent applications, is presented in a manner which assumes only the haziest recollection of elementary fluid-dynamic principles.

Although reference is made throughout to primary source material the approach in this respect, as in others, is a focused one, no attempt having been made to mix into the narrative a comprehensive survey of the background literature; to have done so would have resulted in a very different book from the one intended.

In deciding on how much detail to include I have been guided by experience in presenting much of this material in Master's level degree courses in Italy and the UK. Students on the whole have no problem with being reminded of simple standard procedures, and a number positively welcome it; I have extrapolated these responses to the anticipated readership. In order to avoid clutter some common manipulations are given in small-type paragraphs, which may be easily skipped over. In this way I hope to have defused objections to having, say, spelt out the steps in the formulation of a differential equation from a control volume balance, or the subsequent linearization procedure. Such criticism as may remain in that respect I feel can be lived with. What I have strenuously tried to avoid is the all too familiar cry for help from careful readers of the scientific literature: where on earth does that come from?

The analyses presented in this book represent, by and large, a body of research that has appeared in numerous publications (predominantly in the chemical engineering literature) – some quite recent, others going back over nearly 20 years. In gathering these together for the purpose of producing a coherent narrative I have taken the opportunity to re-order much of the material, to correct errors and inconsistencies, and to add the details and clarifications that space constraints prohibit in journal publications. The book could form the basis for university course modules in engineering and applied science at both undergraduate and graduate level, as well as for focused post-experience courses for the process and allied industries.

L.G. Gibilaro

Acknowledgements

Many people have contributed to the continuing programme of research on which this book draws. First and foremost is Pier Ugo Foscolo, without whom there would be no question of the work having got off the ground. His was the driving force which turned what for me would probably have been just a passing curiosity into a positive crusade. His insight, analytical skill and patient probing of the research literature eventually uncovered the elements of an accessible theory waiting to be assembled. We have worked together throughout the developments described in this book.

That our initial focus was the forces acting on individual fluidized particles is an indication of the influence of Peter Rowe, who provided facilities, advice and constant encouragement throughout the early stages of this work in the Chemical Engineering Department laboratories at the University College London. He had long recognized the importance of such interactions and had subjected them to pioneering experimental study, of direct relevance to the present programme, many years earlier at the United Kingdom Atomic Energy Authority research laboratories at Harwell.

One of the undoubted satisfactions of academic life is that of witnessing the progress of certain research students from eager beginners, struggling to make some sort of sense of the ill-defined open-ended problems they have been handed, to polished professionals, patiently explaining in simple terms to their advisors the steps taken in arriving at momentous conclusions. It does not happen all that often, but three clear instances in the course of this programme call for emphatic acknowledgement: Renzo Di Felice, now Professor of Chemical Engineering at the University of Genova, and an established international authority on liquid fluidization, for his continuing active participation in experimental and theoretical aspects of the work too numerous to individualize; Stefano Brandani, formerly of the University of L'Aquila, now Reader in Chemical Engineering at University College London, for initiating the 'jump condition' analysis reported in Chapter 14 (his doctoral research was supervised by the late Gianni Astarita, whose inspired contributions to this and other areas of the work are also gratefully acknowledged); and Zumao Chen, who as my doctoral student at L'Aquila worked on aspects of slugging fluidization described in Chapter 15, and who subsequently, on his own initiative, embarked on the two-dimensional numerical simulation studies reported in Chapter 16, which have now come to represent the starting point for new programmes of computational research.

From its inception, the work has involved close collaboration between the fluidization research teams of L'Aquila and UCL, accompanied by shuffling of staff and exchange of students. This remains as strong as ever thanks to the active participation of John Yates, who heads the UCL team. Past members of that group who deserve special mention for their contributions to the initial stages of the work are Simon Waldram and Ijaz Hossain; a more recent addition to the team is Paola Lettieri, who maintains strong connections with L'Aquila; David Cheeseman provides continuity and experimental expertise. Major contributors from L'Aquila include Sergio Rapagnà, now Professor of Chemical Engineering at the University of Teramo, Nader Jand and Paolo Antonelli.

I am especially grateful to Yuri Sergeev, Professor of Engineering Mathematics at the University of Newcastle, for his contribution to the 'jump condition' studies, and also for his frequently solicited advice on technical problems encountered along the way. His careful reading of the original draft manuscript resulted in many suggestions for improvement, all of which have been adopted.

Finally, the man who laid the foundation from which we were able to build: Graham Wallis. His unpublished 1962 paper, which he sent me following the appearance of our early applications of his stability criterion, contains a wealth of insight and analysis that, together with his book *One-Dimensional Two-Phase Flow*, we have drawn on repeatedly throughout the course of this work. His direct participation in the programme, at UCL in 1990, provided an invaluable opportunity to clarify aspects of the theory, and to repeat the key 'raining-down' experiments for measuring dynamic wave speed, which he devised and first performed in Peter Rowe's laboratory in Harwell, and only reported in the unpublished 1962 paper; the method is described in Chapter 9, along with his original results and our more recent ones.

Research origins

This book gives an account of a formulation of the conservation equations for mass and momentum in a fluidized suspension, and applications of these equations to the prediction of system behaviour. The history of this approach to fluidization research is relatively recent, and the brief account which now follows is based largely on the personal recollections of the major players, recounted to me in private conversations and correspondence.

It all started in 1959. Robert Pigford, a professor of chemical engineering from the University of Delaware in the USA, was taking sabbatical leave in England, at the University of Cambridge. It was there that he undertook what appears to have been the first formal analysis of the stability of the fluidized state, arriving at an unexpected and far-reaching conclusion, which was eventually to be embraced by the academic community. This acceptance came much later, however, and as a result of the same conclusion being reached and published by somebody else – Roy Jackson, then of the University of Edinburgh. His account appeared in 1963, 4 years after Pigford's original discovery, which eventually surfaced in the scientific literature, with modifications and a co-author, some 2 years later. I learned the story behind this long delay in publication in 1987. It was recounted to me over a lengthy lunch in the Faculty Club of the University of Delaware. Robert Pigford had retired by then, but was still supervising research. He died some months after this, our only encounter.

In 1959 he found himself in the Chemical Engineering Department at Cambridge with time on his hands, and so took the opportunity to attend a course of lectures on fluid mechanics, a topic quite new to him, given by a brilliant applied mathematician by the name of George Batchelor. He found this immensely stimulating, opening his eyes to all manner of applications to problems he had hitherto regarded as quite intractable – in particular, to the question of why powders fluidized by gases behave in a highly disturbed, vigorously agitated manner, far removed from the ordered state of uniform suspension that intuition and simple theoretical considerations would at first suggest. Such systems had recently come to occupy positions of prime importance in the process industries, and as a consequence had become the subject of extensive empirical study.

He lost no time in putting his newly acquired skills into effect, arriving after some labour at a remarkably satisfying conclusion: his seemingly general mathematical description of a uniformly fluidized bed showed it to be *intrinsically unstable*; tiny imposed perturbations, he found, would grow at phenomenal rates, leading to precisely the physical manifestations that had previously defied rational explanation. Nearly 30 years later, he still regarded this discovery as the most significant achievement in his long and exceptionally distinguished career in academic chemical engineering. He wrote it up and submitted it to Batchelor, his mentor and source of inspiration, who 3 years earlier had founded (and since edited) the prestigious *Journal of Fluid Mechanics*. Such was his excitement that he expected an enthusiastic response within days. But the days turned to weeks; and it was only after he had all but given up hope of ever receiving a reply that there appeared in his post tray an envelope containing a brief handwritten note of summary rejection. He returned to the USA in low spirits, eventually to publish, but only after his key conclusion had already been accorded the status of an established truth.

The second episode in the unfolding saga bears some similarity to the first. In 1962 a paper on fluidization-dynamics, of prophetic importance as it turned out, was also submitted to the *Journal of Fluid Mechanics*, this time by Graham Wallis – a name soon to become widely associated with seminal advances in the field of two-phase flow. His analysis was based on a particle momentum equation which included a term in addition to those appearing in the formulation proposed by Pigford 3 years before. With this extra term, the model was able to describe both stable and unstable fluidization, and to predict the possibility of a bed switching between these two states under certain conditions of operation. The paper was never published. More than 10 years were to pass before observations of this predicted behaviour in gas-fluidized beds were to be reported in the literature.

To the world at large, quite unaware of these suppressed and somewhat conflicting revelations, the first inkling of the problem surfaced in 1963 with the paper by Jackson referred to earlier. As already indicated, his major conclusion – that the state of uniform particle suspension is intrinsically unstable – coincided with Pigford's. During the course of the reviewing procedure it seems he was made aware of this previous work, copies of which had found their way into the hands of Cambridge colleagues. (I am grateful to John Davidson of the Chemical Engineering Department at Cambridge for sending me such a copy, containing Pigford's handwritten corrections.) Jackson refers to this unpublished work in his paper, drawing some comfort from the fact that it contained an inconsistency in the way in which the term describing the interaction of the fluid with the particles was formulated. Readers who persist with this book will soon come to learn the reason why the key conclusion on which they both converged is effectively independent of the formulation of this primary interaction term.

This insensitivity to details of the basic Pigford/Jackson model was to have a profound effect in cementing views on the nature of the fluidization process. It seemed that however the interaction between fluid and particles is described, the essential conclusion remains unchanged: the uniform fluidized state remains intrinsically unstable. So when irrefutable experimental evidence for *stable* gas fluidization became available in the mid-1970s, the initial reaction was one of disbelief, soon to be followed by an earnest search for a way out of the dilemma. As far as the fluidization research community was concerned, the additional term in Wallis's formulation, which predicted just such an outcome, remained in the shadows – despite its appearance in his 1969 book, *One-Dimensional Two-Phase Flow*, and the promising results of a specific application of his general conclusion by Dutch researchers Verloop and Heertjes soon after.

The need for the extra term in the momentum equation was eventually taken on board, but only when it had been rediscovered by others (a seemingly persistent theme in the story). However, this time, in what amounted to a very convenient compromise, bringing relief and comfort to almost all the participants, a novel interpretation was proposed. Instead of describing just another aspect of fluid dynamic interaction, Wallis's additional term was attributed to a quite different mechanism: the sticking together of the particles, by means of 'particle–particle' contact forces, to form a coherent expanded structure. In this way the uniform state of suspension was allowed to remain unstable in purely fluid dynamic terms – in full accord with the Pigford/Jackson formulation. However, then this other mechanism, of essentially unquantifiable nature, was deemed to come into play, heralding the intense programmes of experimental investigation which were soon to follow.

It was against the backdrop of these vicissitudes that the analysis of the fluidized state that is the subject of this book was to take shape. This account of the origins of the research now becomes strictly personal, commencing with the inauspicious beginnings of a new initiative, which, in keeping with the enshrined tradition, hinged on the rediscovery of a long established relation.

I had been asked by Peter Rowe, who headed the Chemical Engineering Department at University College London, to advise on some peripheral aspect of a manuscript submitted to him in his capacity as editor of a scientific journal. All I now remember of the work is that it related to liquid-fluidized beds subjected to changes in liquid flow rate; and that it seemed to imply, as I understood it at the time, that a quantitative descriptive mechanism for this behaviour was unavailable. As a result of this perceived deficiency, I and a colleague, Simon Waldram, spent the next week or so trying to model the appropriate transient response of the surface elevation of a fluidized bed, eventually coming up with a result of breathtaking simplicity (reported in Chapter 5 of this book). Experiments performed on a hastily constructed experimental rig confirmed the essential predictions of the model, precipitating scenes of self-congratulatory revelry. Then, the morning after, a belated examination of the literature revealed that the same conclusion had been published some 20 years previously. The fact that this earlier analysis followed a quite different route from ours provided scant consolation at the time.

This episode would probably have marked the end, as well as the beginning, of my incursion into the realm of fluidization research were it not for the arrival on the scene of Pier Ugo Foscolo. I had previously resisted invitations to become involved in this field, largely on the basis that, as so many formidably gifted persons had laboured in it for so long, the remaining pickings were likely to be meagre, if not totally inaccessible. Foscolo, on the other hand, had no such inhibitions. He had arranged sabbatical leave from the University of L'Aquila in Italy to work in Rowe's group at UCL, initially for 1 year, later, in view of developments described in the early chapters of this book, to be extended to 2 years. This marked the start of a research collaboration that continues to this day, reinforced in the autumn of 1993 by my transfer from UCL to the University of L'Aquila.

The problem with which Foscolo was wrestling at the time (in part as an escape from the tedium of the experimental programme, involving precise measurements of X-ray photographs of bubbles in gas-fluidized beds, which justified his appointment at UCL) was closely related to the one that had given rise to the mood swings described earlier. It concerned the equilibrium characteristics of liquid-fluidized beds, which are known to obey a remarkably simple *empirical* law for which no rational explanation was forthcoming. In view of the immense expenditure of intellectual and manual effort directed at essentially complex aspects of fluidized bed behaviour, it appeared strange at the time (and still does today) that this simple relation had remained largely exempt from analytical consideration. The outcome of this investigation is described in Chapters 3 and 4. In addition to establishing a clear link between fluidized bed expansion and the mechanism of fluid flow through porous media, the analysis was to lead to compact, fully predictive expressions for the primary forces acting on a fluidized particle; these were to play a major role in subsequent developments.

A significant breakthrough was soon to follow. It involved an explicit formulation of Wallis's fluid-dynamic criterion for the stability of the homogeneously fluidized state. Our formulation (described in Chapter 6) drew heavily on the two initial investigations referred to above, together with Foscolo's inspired innovation of treating the suspension of fluidized particles as formally analogous to a compressible fluid. This gave rise to a simple algebraic expression, requiring solely a knowledge of the basic fluid and particle properties, which provided an immediate answer to the question of whether the fluidization would be *stable* or *unstable* for any specified system. These two regimes are generally associated with liquid and gas fluidization respectively. The criterion was able to distinguish quite unambiguously between these two markedly different manifestations of the fluidized state, and to identify those intermediate systems that, at a clearly defined fluid flow rate, switch from stable to unstable behaviour.

The chapters that follow give an account of a simple fluid-dynamic theory for the fluidized state. At its heart lies a specific formulation of Wallis's additional term in the particle momentum equation, describing a fluid-dynamic mechanism whereby the suspended particle assembly comes to adopt effectively elastic properties. The full formulation will be seen to lead to quantitative predictions of many aspects of fluidized bed behaviour, a feature that is emphasized throughout the book by means of direct comparison of model solutions with experimental observations.

Notation

a	perturbation amplitude growth rate, s^{-1}
a	exponent of fluid flux
A	particle projected area, m^2
Ar	Archimedes number (defined by eqn (2.14))
b	exponent of void fraction
B	defined in eqn (7.16), s^{-1}
B_p	bulk mobility of the particles, s/kg
C	defined in eqn (7.16), m/s^2
C_D	unhindered particle drag coefficient
$C_{D\alpha}$	fluidized particle drag coefficient
d_p	particle diameter, m
D	tube/bed diameter, m
D	defined by eqn (8.29), s^{-1}
D_e	effective tube diameter, m
De	Density number
E	elastic modulus of particle phase, N/m^2
f	friction factor
f_{u_p}	defined by eqn (7.14), Ns/m^4
f_{ε}	defined by eqn (7.14), N/m^3
f	particle net primary force, N
f	bed surface oscillation frequency, s^{-1}
f_0	particle net primary force at equilibrium, N
f_b	particle buoyancy force, N
f_d	particle drag force, N
f_g	particle gravitational force, N
f_I	particle interaction force, N
$f_{\Delta z}$	particle elastic force, N

f^+ particle net force, N
F net primary force, N/m^3
F_b buoyancy force, N/m^3
F_d drag force, N/m^3
F_I fluid particle interaction force, N/m^3
F^+ net force, N/m^3
Fl Flow number
Fr Froud number
g gravitational field strength, N/kg
g_s simulated gravitational field strength, N/kg
G defined by eqn (11.8), m^2/s^2
Ga Galileo number
H bed height, m
H_o initial bed height, m
k wave number, m^{-1}
K_D Darcy equation constant, 1/m^2
l length element, m
L length element, m
L_A bed surface displacement, m
L_B bed length, m
L_e effective length, m
L_l lower zone length, m
L_{SS} solid slug length, m
Le Length number
Mo Mobility number
n Richardson–Zaki exponent
N_L number of particles per unit area, m^{-2}
N_V number of particles per unit volume, m^{-3}
p fluid pressure, N/m^2
$\widehat{p}$ root mean square pressure, N/m^2
p_p particle pressure, N/m^2
P defined by eqn (11.21)
Re_p particle Reynolds number
Re_t particle terminal Reynolds number
S stability function
t time, s
T tortuosity
T_T transient response time, s
u_{bs} bed surface velocity, m/s
u_D dynamic wave velocity, m/s

u_{DT} two-phase dynamic wave velocity, m/s
u_{DS} dynamic shock velocity, m/s
u_f fluid velocity, m/s
u_{f0} fluid velocity at equilibrium, m/s
u_{fp} relative fluid particle velocity, m/s
u_{FS} fluid slug velocity, m/s
u_{K} kinematic wave velocity, m/s
u_{KS} kinematic shock velocity, m/s
u_p particle velocity, m/s
u_{p0} particle velocity at equilibrium, m/s
u_t unhindered particle settling velocity, m/s
U volumetric flux, m/s
U_0 volumetric flux at equilibrium, m/s
U_A upper zone flux, m/s
U_{mb} minimum bubbling flux, m/s
U_{mf} minimum fluidization flux, m/s
v wave velocity, m/s
v_f lateral fluid velocity, m/s
v_p lateral particle velocity, m/s
V shock velocity, m/s
V defined by eqn (11.9), m/s
V_{D} defined by eqn (10.2), m/s
V_B volume of bed particles per unit of cross-section, m
V_p particle volume, m^3
w_e particle effective weight, N
x lateral distance, m
z axial distance, m
α particle volumetric concentration
α_e effective particle volumetric concentration
γ particle concentration ratio, eqn (14.19)
δE energy dissipation rate, W/m^2
δE_d energy dissipation rate, W/m^2
δE_{KE} energy dissipation rate due to kinetic energy creation, W/m^2
δE_{PE} energy dissipation rate due to potential energy creation, W/m^2
Δa fluidization quality parameter, eqn (10.7), s^{-1}
$\Delta \mathrm{P}$ unrecoverable pressure-loss, N/m^2
$\Delta \mathrm{P_B}$ unrecoverable pressure loss across entire bed, N/m^2
$\Delta \mathrm{P}_{KE}$ pressure loss due to kinetic energy creation, W/m^2
$\Delta \mathrm{P}_{PE}$ pressure loss due to potential energy creation, W/m^2
ε void fraction

ε_0 equilibrium void fraction
ε^* void fraction deviation
ε_A initial perturbation wave amplitude
ε_d dense phase void fraction
ε_{df} void fraction at dilute fluidization regime boundary
ε_{dn} void fraction at jump 'nose'
$\varepsilon_{d\infty}$ dense phase void fraction at high fluid flux
ε_{mb} void fraction at minimum bubbling condition
ε_{mf} void fraction at minimum fluidization condition
θ particle layer spacing, m
φ angle of internal friction
λ wave length, m
μ_f fluid viscosity, Ns/m^2
ρ_f fluid density, kg/m^3
ρ_p particle density, kg/m^3
ρ_{pp} particle phase density, kg/m^3
ρ_s suspension density, kg/m^3

subscripts
1, 2 before and after shock front
x, z lateral and axial directions

superscripts
$\hat{x}$ dimensionless scaled value of quantity x
$\breve{x}$ quantity x relative to datum value

bold type
$\mathbf{x}$ vector quantities

1

Introduction: the fluidized state

Fluidization

Fluidization is a process whereby a bed of solid particles is transformed into something closely resembling a liquid. This is achieved by pumping a fluid, either a gas or a liquid, upwards through the bed at a rate that is sufficient to exert a force on the particles that exactly counteracts their weight; in this way, instead of a rigid structure held in place by means of gravity-derived contact forces, the bed acquires fluid-like properties, free to flow and deform, with the particles able to move relatively freely with respect to one another.

A number of colourful demonstrations have been devised to illustrate this transformation. One that for many years occupied a prime position in the Chemical Engineering Department laboratories at University College London, later to appear at the Science Museum in Kensington, involved a bed of fine sand and, among other artefacts, two toy ducks, one plastic and one brass. The low-density plastic duck is buried deep in the sand and the high-density brass one is placed on the surface; a compressed air supply to the bottom of the bed is then turned on and the flow progressively increased. When the fluidization point is reached the brass duck sinks to the bottom and the plastic one pops to the surface, where it floats about just as it would in water.

The same principle can be observed in another demonstration, which serves a practical as well as an heuristic purpose. This time, salt crystals are fluidized with air in a container fitted with an electric immersion heater. For reasons that will be discussed in the following chapters, the beds described in both this and the previous example come to resemble a *boiling* liquid at air flow rates above that required to just fluidize the bed; bubbles of air rise rapidly through the fluidized particles causing vigorous mixing, and then burst through the surface – just like steam bubbles in boiling water. This mixing, induced by the bubbles, ensures that the whole bed acquires a uniform temperature. The demonstration now involves dropping corn grains on to the bed surface; their density is a little greater than that of the hot salt suspension, so they sink initially, then heat up and 'pop'. The low-density popcorn immediately rises to the surface, ready salted, for collection and consumption.

This second demonstration illustrates a number of useful features of the fluidized state as a processing environment. In addition to the obvious advantages resulting from the acquisition by the particles of fluid-like properties, which permit them to flow freely from one location to another, the high level of particle mixing means that heat and mass can be rapidly transferred throughout the bed, with far-reaching consequences for its performance as a chemical reactor.

Applications

A major application of fluidized bed technology is to be found in the catalytic-cracking reactor, or 'Cat Cracker', which lies at the heart of the petroleum refining process. Here, the catalyst particles (which promote the breakdown of the large crude petroleum molecules into the smaller constituents of gasoline, diesel, fuel oil, etc.) are fluidized by the vaporized crude oil. An unwanted by-product of the reactions is carbon, which deposits on the particle surfaces, thereby blocking their catalytic action. The properties of the fluidized state are further exploited to overcome this problem. The catalyst is reactivated continuously by circulating it to another bed, where it is fluidized with air in which the carbon burns away, and then back again, regenerated, to continue performing its allotted catalytic function.

Other applications, established and potential, are boundless. Gas-fluidized beds are widely used as chemical reactors, and also as combustors to raise steam for power generation. This latter application can involve the burning of coal, and both urban and agricultural waste, in air-fluidized sand beds. Agricultural waste and purpose-grown energy crops can be fluidized in steam to produce a hydrogen-rich fuel gas. Liquid-fluidized beds are employed extensively in water treatment, minerals processing and fermentation technology.

Research

Research into the mechanisms of the fluidization process falls largely into two distinct categories: applied research, involving actual process plant or, more usually, laboratory units that seek to mimic the particular feature of the process plant that is the subject of study; and theoretical analysis, rooted in the rigorous framework of multiphase fluid mechanics. The former is the province of the engineer, the latter of the mathematician. Although instances of cross-fertilization have been known, such occurrences are rare. The theoretician who strays into the factory is appalled at the physical imponderables that characterize the real world, as is the engineer by the mathematical complexities in the analysis of a supposedly physical problem, even when it has been so simplified at the onset as to render it totally inapplicable to any conceivable practical application.

The analysis reported in this book is representative of a middle way that seeks to model the essential features of the fluidized state by imbedding in the basic theoretical framework (the conservation laws for mass and momentum) simple formulations of the primary force interactions, and drawing on formal analogies with theoretical treatments of simpler, well-posed physical problems possessing the same mathematical structure.

Single particle suspension

An obvious starting point for the examination of the mechanism of the fluidization process, which involves the suspension of a very large number of solid particles in an upwardly flowing fluid, is the much simpler case of the single particle.

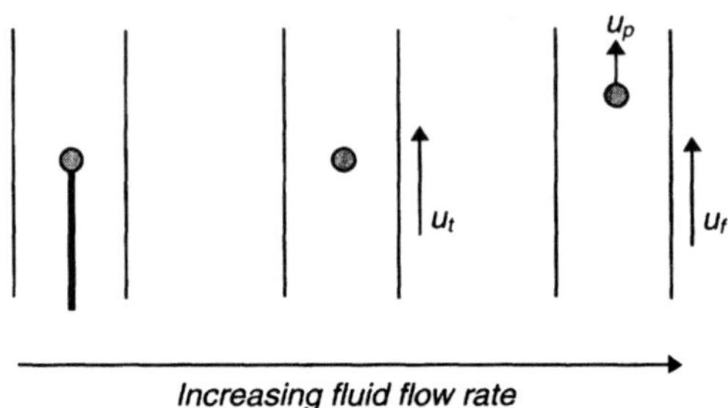

Figure 1.1 Single particle suspension and transport.

Consider a solid sphere sitting on a small support in a vertical tube (Figure 1.1, left). A fluid (either gas or liquid) is pumped up the tube so that it imparts an upward force on the sphere. As the fluid flow is progressively increased, this upward force reaches the critical value (at fluid velocity u_t) that just balances the sphere's weight; at this point the support structure can be removed and the sphere will remain stationary, supported entirely by the force of interaction with the fluid stream (Figure 1.1, middle).

If the fluid flow rate is now increased beyond this critical value u_t, the magnitude of the interaction force becomes greater than that due to gravity, giving rise to a net force that causes the sphere to accelerate upwards. As it does so, its velocity relative to the fluid (and, as a consequence, the interaction force) decreases progressively until it reaches the critical value at which the gravitational force is again just balanced:

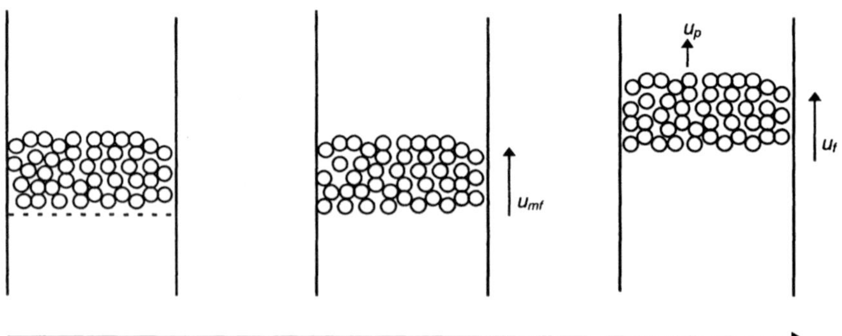

Figure 1.2 The minimum fluidization point.

$u_f - u_p = u_t$. From this point on the sphere continues its upward motion in equilibrium, at constant velocity u_p: $u_p = u_f - u_t$ (Figure 1.1, right).

We can try to apply these simple considerations, relating to a single solid sphere, to a bed consisting of a large number of such spheres supported on a mesh that extends over the entire tube cross-section (Figure 1.2). A progressive increase in the fluid flow rate will, once again, lead to the critical condition at which the total weight of the particles is just balanced by the fluid–particle interaction force (the *minimum fluidization* condition); at this point it could be thought possible to dispense with the supporting mesh, leaving the particle bed suspended motionless in the fluid stream, as was the case for the single sphere.

Continuing with the reasoning, we might expect a further increase in the fluid flow rate to cause the assembly of particles to accelerate upwards together, until such time as the relative velocity of the fluid ($u_f - u_p$) has fallen to that of the critical, minimum fluidization condition and equilibrium is re-established; from this point on the particle assembly would proceed up the tube, piston-like, at constant velocity.

Such behaviour, following the minimum fluidization point, does not occur in practice unless the particles are glued together. What precisely does happen is described in some detail in the following chapters, and depends on the properties of the particles and fluid involved. We will see that one possibility, commonly encountered when the fluidizing agent is a liquid, is that the bed 'expands' to an essentially homogeneous condition in which the particles are separated from one another more or less uniformly, with relatively little particle motion, the extent of the separation increasing progressively with increasing fluid velocity. Another possibility, more usually encountered with gas-fluidized systems, has already been mentioned: this time all the fluid in excess of that required to just bring the particles to the minimum fluidization point forms rising bubbles, which cause considerable particle mixing and give the bed the appearance of a boiling liquid. Various terms have been adopted to describe these two quite different manifestations of the fluidized state. We shall refer to them as *homogeneous* and *bubbling* fluidization respectively.

Fluidization quality

Homogeneous and bubbling fluidization represent two quite different fluid-dynamic environments brought about by the fluidization process itself. They may be regarded as somewhat extreme examples of *fluidization*

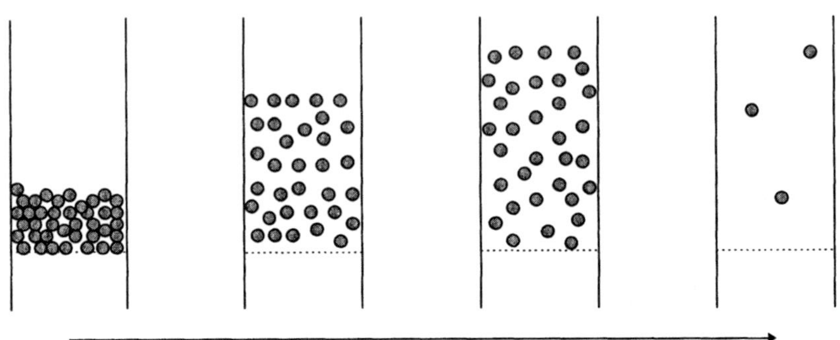

Figure 1.3 Homogeneous fluidization – from packed bed to single particle suspension.

quality, a conveniently vague term that nevertheless serves to portray the fluidized state as a continuous spectrum of behavioural conditions. Given that the main applications of fluidized bed technology rely on the provision of intimate contact between the solid and fluid phases for the purpose of promoting chemical reactions, it is hardly surprising that fluidization quality is a key factor in determining the performance of a fluidized bed as a chemical reactor. A major incentive for the analyses reported in the following chapters has been the urgent need for means of quantifying the essential factors that determine fluidization quality; and for predicting, on the basis of the particle and fluid properties and conditions of operation, the fluidization quality that would result in an envisaged fluidized bed reactor.

Homogeneous fluidization

The conceptually simplest means by which particles can remain in a bed, subjected to a fluid flux higher than that required for minimum fluidization, is for them to separate from one another so that the bed expands, the void space around the particles increases and, as a consequence, the fluid velocity within the bed decreases. This decrease in *interstitial velocity* has a strong effect on the fluid–particle interaction force, causing it to fall and thereby enabling a new equilibrium condition to be established in which the particle weight is once again just supported by the fluid. The mechanism just described represents the essential feature of *homogeneous fluidization.*

At first sight, there would appear to be no reason why any fluidized system should not be operated homogeneously at any fluid flow rate within the range bounded by the minimum fluidization velocity on the one hand and the velocity required to just support a single particle in the otherwise empty tube on the other. An equilibrium condition can always be identified within this range, but, as we shall see, other criteria must be satisfied in order for this condition to be attainable in practice. These considerations are best delayed until after the state of equilibrium itself has been examined.

Any analysis of the homogeneously fluidized state must encompass the conditions of single particle suspension and fluid flow through fixed beds of particles; these represent, respectively, the upper and lower bounds for fluidization as illustrated in Figure 1.3.

We start our examination of the fluidized state with brief accounts of established treatments of these upper and lower bounds. Both of these areas have been the subject of copious study, from which we select only those elements that are of direct relevance to the analysis that follows.

Single particle suspension

The single particle settling velocity

A key parameter in the analysis of the fluidized state turns out to be the *unhindered terminal settling velocity* (u_t) of a single particle in the stagnant fluidizing medium. For the case of a liquid, u_t may be easily measured by releasing the particle at the surface of a transparent vessel containing the liquid, and timing its passage between two reference levels situated sufficiently below the surface to ensure that the *terminal*, constant velocity condition has been reached; the vessel diameter must also be sufficiently large with respect to the particle for the *unhindered* condition to apply.

The equilibrium condition experienced by a particle falling at velocity u_t in a stationary fluid is, of course, equivalent to that of a motionless particle suspended in an upwardly flowing fluid with velocity u_t: this latter situation represents the upper fluid velocity bound for homogeneous fluidization.

The forces acting on the particle are the result of fluid–particle interaction f_I and gravity f_g. Under conditions of equilibrium, we have:

$$f_I + f_g = 0. \tag{2.1}$$

For the case of essentially spherical particles and Newtonian fluids, u_t can be readily estimated from this relation over the entire flow regime of relevance to the fluidization process.

The creeping flow regime

A rigorous solution exists for f_I for the limiting condition of very low fluid flow rates around a sphere – in which the fluid streamlines follow the contours of the sphere, with no separation at the upper surface (the so called *creeping flow* regime). This may be regarded to occur at particle Reynolds numbers Re_p below about 0.1:

$$Re_p < 0.1, \qquad Re_p = \frac{d_p u_f \rho_f}{\mu_f}. \tag{2.2}$$

Many industrial fluidized bed reactors operate within, or close to, this range.

The total fluid–particle interaction force f_I can be obtained from the integral over the entire particle surface of local, point interactions – for a detailed derivation see, for example, Bird *et al.* (1960). This operation gives rise to the remarkably compact form:

$$f_I = \frac{\pi d_p^3}{6} \rho_f g + 3\pi d_p \mu_f u_f, \tag{2.3}$$

where the first term will be seen to represent the Archimedean *buoyancy force* f_b – the net effect on the particle of the pressure gradient in the fluid itself – and the second term describes the total *drag force* f_d, which is a consequence of energy dissipation at the particle surface, and is proportional to the mean velocity of the fluid relative to that of the particle:

$$f_I = f_b + f_d. \tag{2.4}$$

The buoyancy force f_b is independent of the fluid flow regime, whereas for higher Reynolds number conditions the drag force f_d becomes a nonlinear function of the relative fluid–particle velocity, for which empirical correlations are generally required (although a solution does exists for higher Reynolds number laminar flow: Proudman and Pearson, 1956). This is a well-trodden path that has yielded many, more or less equivalent, expressions from which f_d may be estimated.

The drag coefficient

Empirical relations are best expressed in dimensionless form. For the case of a sphere in a fluid stream, the drag force is made dimensionless by dividing by any convenient reference level that also possesses the dimensions of force. Thus the dimensionless drag force, or *drag coefficient* C_D, may be expressed:

$$C_D = \frac{f_d}{(\rho_f u_f^2/2)(\pi d_p^2/4)}; \tag{2.5}$$

where the denominator, the chosen reference level, is the product of the kinetic energy possessed by a unit volume of the fluid and the projected area of the sphere. Although quite arbitrary, this has become the standard definition of the drag coefficient.

The creeping flow regime

On substituting into eqn (2.5) the expression for the creeping flow regime drag force, $f_d = 3\pi d_p \mu_f u_f$, we obtain:

$$C_D = \frac{24\mu_f}{u_f \rho_f d_p} = \frac{24}{Re_p}. \quad \textit{Creeping flow regime} \tag{2.6}$$

The inertial flow regime

Values of Re_p above about 500 represent the inertial flow regime, for which the drag coefficient has been found to be approximately constant:

$$C_D \approx 0.44. \quad \textit{Inertial flow regime} \tag{2.7}$$

(At very high particle Reynolds numbers ($> 10^5$), C_D is found to fall sharply as a result of a sudden shift in the boundary layer separation zone. This condition is well outside the range of relevance for fluidization.)

All flow regimes

Copious experimentation has confirmed the validity of the above limiting forms, and led to correlations for the intermediate flow regime in terms of the particle Reynolds number. A compact form, which for most practical purposes adequately represents all the data, is attributed to Dallavalle (1948):

$$C_D = (0.63 + 4.8Re_p^{-0.5})^2. \quad \textit{All flow regimes} \tag{2.8}$$

This relation is shown in Figure 2.1, together with the creeping flow and inertial limits, equations (2.6) and (2.7) respectively.

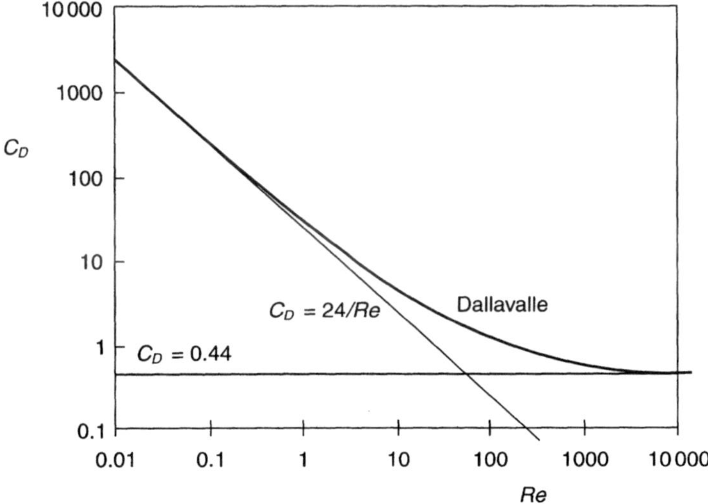

Figure 2.1 Drag coefficient as a function of particle Reynolds number.

The terminal velocity u_t

It is convenient to refer to the net effect of gravity and buoyancy on a particle (Figure 2.2) as the *effective weight* w_e:

$$w_e = f_g + f_b = -\frac{\pi d_p^3}{6}(\rho_p - \rho_f)g. \tag{2.9}$$

Under terminal, equilibrium conditions ($u_f = u_t$, $Re_p = Re_t$), the drag force equates to this effective weight,

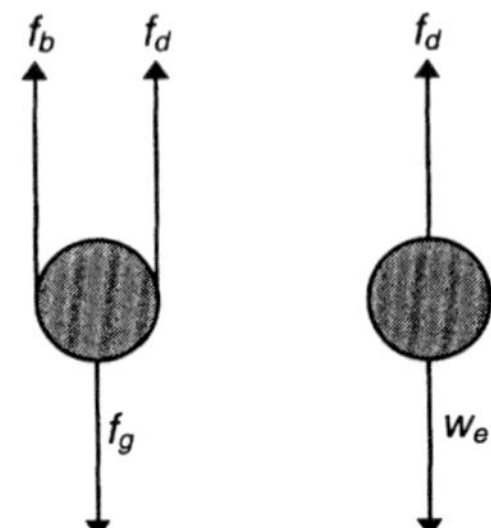

Figure 2.2 Primary single particle forces.

$$f_d = -w_e, \tag{2.10}$$

a relation which enables the drag coefficient to be related to u_t:

$$C_D = \frac{4}{3} \cdot \frac{gd_p}{u_t^2} \cdot \frac{(\rho_p - \rho_f)}{\rho_f}. \tag{2.11}$$

As C_D is itself a function of Re_t, and hence of u_t, this equation can always be solved iteratively for u_t. The expressions for C_D quoted above, however, lead to explicit forms:

$$u_t = \frac{(\rho_p - \rho_f)gd_p^2}{18\mu_f}. \quad \textit{Creeping flow regime} \tag{2.12}$$

This fully theoretical expression for u_t in the creeping flow regime is known as *Stokes Law*.

$$u_t = \sqrt{3.03gd_p(\rho_p - \rho_f)/\rho_f}. \quad \textit{Inertial flow regime} \tag{2.13}$$

Dimensionless relations

The above expressions for the terminal velocity of a single particle may be expressed in dimensionless form, thereby introducing another of the dimensionless groups, the *Archimedes* number *Ar*, which will subsequently be used in the characterization of the fluidized state.

$$Re_t = \frac{d_p u_t \rho_f}{\mu_f}, \qquad Ar = \frac{gd_p^3 \rho_f(\rho_p - \rho_f)}{\mu_f^2}. \tag{2.14}$$

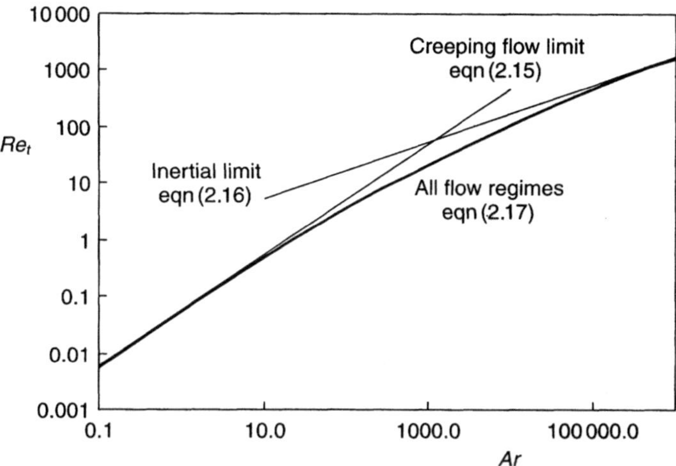

Figure 2.3 Terminal Reynolds number as a function of Archimedes number.

In terms of these groups, the above terminal settling velocity relations become:

$$Re_t = \frac{Ar}{18}; \quad \textit{Creeping flow regime} \tag{2.15}$$

$$Re_t = \sqrt{3.03Ar}; \quad \textit{Inertial flow regime} \tag{2.16}$$

$$Re_t = \left[-3.809 + (3.809^2 + 1.832Ar^{0.5})^{0.5}\right]^2. \quad \textit{All flow regimes} \tag{2.17}$$

This final, general expression, based on the Dallavalle correlation, eqn (2.8), enables Re_t, and hence u_t, to be estimated for any system for which the particle diameter and density and the fluid density and viscosity are known. It is illustrated in Figure 2.3.

References

Bird, R.B., Stewart, W.E. and Lightfoot, E.N. (1960). *Transport Phenomena*. Wiley.

Dallavalle, J.M. (1948). *Micromeritics*. Pitman.

Proudman, I. and Pearson, J.R.A. (1956). Expansions at small Reynolds numbers for the flow past a sphere and a circular cylinder. *J. Fluid Mech.*, **2**, 237.

3

Fluid flow through particle beds

Fluid pressure loss in packed particle beds

The upper fluid velocity limit for fluidization, u_t, was examined in the previous chapter. We now turn to the lower limit, below which the particles are stationary and in direct contact with their neighbours. Under these conditions the interaction force is insufficient to support the weight of the particles; all that happens is that the fluid, as it rises through the bed, loses energy due to frictional dissipation, resulting in a loss of pressure that is greater than can be accounted for by the progressive increase in gravitational potential energy. It is clearly important to be able to estimate this additional energy requirement, and considerable research effort has been expended for this purpose. We consider first the reasoning behind the most widespread of the methods adopted, and then go on to consider the modifications that become necessary to make it applicable to the fluidization process.

The unrecoverable pressure loss

The total drop in fluid pressure across a length L of bed is Δp. Of this a portion $\Delta \mathrm{P}$ comes about as a result of frictional interaction between the fluid and the particles; it represents energy irrevocably lost by the fluid, dissipated as heat. It is therefore convenient to refer to $\Delta \mathrm{P}$ as the *unrecoverable pressure loss*. If the total pressure loss in the fluid is attributable solely to fluid–particle frictional interaction and the gain in gravitational potential energy in the rising fluid (as may be assumed in the applications described in this and subsequent chapters), then we have:

$$\Delta \mathrm{P} = \Delta p - \rho_f g L. \tag{3.1}$$

This relation for $\Delta \mathrm{P}$ is more generally portrayed as the definition of the *piezometric* or *manometric* pressure drop: total pressure drop minus the hydrostatic contribution. The loss of generality in the chosen interpretation is compensated for by the clear association of $\Delta \mathrm{P}$ with fluid–particle frictional dissipation in the applications we now consider.

The tube-flow analogy: viscous flow conditions

Theoretical expressions for unrecoverable pressure loss in Newtonian fluids in laminar flow were first derived in the mid-nineteenth century. For a path length L in a cylindrical tube of diameter D, it becomes:

$$\Delta \mathrm{P} = \frac{32}{D^2} \mu_f L U, \quad \textit{The Hagen–Poiseuille equation} \tag{3.2}$$

where μ_f is the fluid viscosity and U the volumetric flux – volumetric flow rate per unit area of tube cross-section. Energy dissipation is in this case brought about by fluid interaction with the tube wall.

At around the same time that Hagen and Poiseuille were (independently) engaged in the theoretical analysis of viscous fluid flow, an experimental investigation was being carried out by a French municipal engineer concerned with the very practical problem of water supply and distribution in urban areas. His experiments involved measuring the permeation rates of water through beds of sand, and led to the empirical relation:

$$\Delta \mathrm{P} = K_D \mu_f L U. \quad \textit{The Darcy equation} \tag{3.3}$$

The similarity in form of these two expressions for $\Delta \mathrm{P}$, eqns (3.2) and (3.3), suggested the use of an analogy with the theoretical Hagen–Poiseuille tube-flow equation for relating the proportionality constant K_D in the Darcy equation to measurable properties of particle beds. This approach has stood the test of time, and remains to this day the most common tactic for estimating ΔP in such systems. The original application of the analogy, which we now outline, involved simply replacing the fluid flux U and tube diameter D in the Hagen–Poiseuille equation with terms relating appropriately to flow through porous media. We will have cause later, when dealing with flow through *expanded beds* (which relate more appropriately to fluidized suspensions), to re-examine the assumptions implicit in the classical treatment.

The fluid velocity

The term U in the Hagen–Poiseuille equation, in addition to representing the volumetric flux of the fluid, may be interpreted as the mean velocity of the fluid relative to the tube wall. The situation is different for the case of the volumetric flux U appearing in the Darcy equation; here a fraction of the bed cross-section is blocked by the particles, leaving only the remaining *void fraction* ε available for flow. Thus a volumetric flux of U corresponds to a relative fluid–particle velocity of U/ε, which provides the first substitution to be made to the Hagen–Poiseuille equation for application to particle beds.

The effective diameter

The geometry of a passage through which a fluid flows determines the flow rate for a specified pressure drop. This flow rate will increase with increasing void volume of the passage, and decrease with increasing wall area (which offers resistance to flow). Thus the ratio of these quantities provides a convenient measure, having the dimension of length, of the permeability of the passage:

$$\textit{Permeability} \propto \frac{\textit{void volume}}{\textit{internal surface area}}. \tag{3.4}$$

For a cylindrical tube of diameter D, this ratio becomes:

$$\textit{Cylindrical tube:} \quad \frac{\textit{void volume}}{\textit{internal surface area}} = \frac{\pi D^2 L/4}{\pi D L} = \frac{D}{4}, \tag{3.5}$$

a result which provides a definition for the *effective diameter* D_e of any flow passage for which the void volume and internal surface can be calculated or measured:

$$D_e = 4 \cdot \frac{\textit{void volume}}{\textit{internal surface area}}, \tag{3.6}$$

so that for a cylindrical tube D_e equates to the tube diameter D.

For a bed of monosize spheres of diameter d_p, a unit of volume contains $6(1-\varepsilon)/\pi d_p^3$ particles, with total surface area $6(1-\varepsilon)/d_p$. Hence:

$$\textit{Monosize spheres:} \quad D_e = \frac{2\varepsilon d_p}{3(1-\varepsilon)}. \tag{3.7}$$

For a bed containing spheres of different sizes, the definition for D_e, eqn (3.6), leads to the same form as the monosize sphere expression, eqn (3.7), if the *surface/volume* average diameter $\bar{d}_p$ is used in place of d_p:

$$\bar{d}_p = \frac{1}{\sum_i \frac{x_i}{d_{pi}}}, \tag{3.8}$$

where x_i is the volume fraction (or mass fraction if the particles all have the same density) of spheres of diameter d_{pi}.

The unrecoverable pressure loss

Making the substitutions for D and U in the Hagen–Poiseuille equation,

$$U \rightarrow U/\varepsilon \quad D \rightarrow D_e, \tag{3.9}$$

yields, for beds of spheres:

$$\Delta \mathrm{P} = 72 \cdot \frac{\mu_f L U}{d_p^2} \cdot \frac{(1-\varepsilon)^2}{\varepsilon^3}. \tag{3.10}$$

The confrontation of this expression for the unrecoverable pressure loss with experimental measurements has led to the constant in eqn (3.10) being increased from 72 to 150, with which value the relation becomes known as the *Blake–Kozeny* equation. A major reason for the increase has been attributed to the fact that fluid flowing through packing follows a tortuous path, which is considerably greater than the bed length L (Carman, 1937). We consider this phenomenon in some detail in the following section, in particular in relation to its effect for expanded particle beds.

The tube-flow analogy: inertial flow conditions

The same procedure described above for low velocity, viscous flow has been applied to the other extreme of high velocity, inertia-dominated flow. In this case the tube-flow equation is expressed in terms of the dimensionless *friction factor* f, which in the inertial flow regime remains essentially constant for a given tube:

$$\Delta \mathrm{P} = 4\mathrm{f} \cdot \frac{L}{D} \cdot \frac{\rho_f U^2}{2}. \tag{3.11}$$

Making the substitutions for U and D as before,

$$U \rightarrow U/\varepsilon \quad D \rightarrow D_e, \tag{3.12}$$

yields:

$$\Delta \mathrm{P} = 3\mathrm{f} \cdot \frac{\rho_f L U^2}{d_p} \cdot \frac{(1-\varepsilon)}{\varepsilon^3}. \tag{3.13}$$

Hardly surprisingly, the experimentally determined value for the constant in the above relation, $3\mathrm{f} = 1.75$, turns out to be orders of magnitude larger than is the case for tube flow. With this value, however, the relation provides a reasonable estimate of many reported observations, and becomes known as the *Burke–Plummer* equation.

The Ergun equation

Simply adding together the expressions for $\Delta \mathrm{P}$ for viscous and inertial conditions yields an equation that has proved capable of providing reasonable estimates of the unrecoverable pressure loss over the whole operating range normally encountered for packed beds. This convenient relation (Ergun and Orning, 1949) is universally referred to as the Ergun equation:

$$\Delta \mathrm{P} = \underbrace{150 \cdot \frac{\mu_f L U}{d_p^2} \cdot \frac{(1-\varepsilon)^2}{\varepsilon^3}}_{\textit{Viscous term}} + \underbrace{1.75 \cdot \frac{\rho_f L U^2}{d_p} \cdot \frac{(1-\varepsilon)}{\varepsilon^3}}_{\textit{Inertial term}}. \tag{3.14}$$

The Ergun equation may also be expressed in terms of the particle Reynolds number $Re_p(=\rho_f d_p U/\mu_f)$:

$$\Delta \mathrm{P} = 1.75 \cdot \frac{L U \mu_f}{d_p^2} \cdot \frac{(1-\varepsilon)}{\varepsilon^3} \cdot \left[85.7\,(1-\varepsilon) + Re_p\right]. \tag{3.15}$$

This form shows that for Re_p equal to 85.7 $(1 - \varepsilon)$, which is approximately 50 for normal packed beds ($\varepsilon \approx 0.4$), the viscous and inertial contributions to the unrecoverable pressure loss are of equal magnitude. For much smaller Re_p the viscous effects clearly dominate, as do the inertial effects for much larger Re_p.

Fluid pressure loss in expanded particle beds

The Ergun equation has been extensively verified for packed beds of spheres and near spheres, for which the void fraction variation remains small: $\varepsilon \approx 0.4$. Some measurements, which we discuss later in this chapter, have been reported for beds artificially expanded by various mechanical means to much higher void fractions; homogeneously fluidized beds can attain void fractions of 0.9 and more. These situations call for a re-examination of the derived dependence of ΔP on void fraction.

The effect of tortuosity

The derivations reported above of the viscous and inertial contributions to the Ergun equation involve the representation of a volume of a porous medium of length L by means of an equivalent cylindrical tube of diameter D_e. The effective length L_e of this tube must clearly be greater than L because of the twisted path followed by the fluid around the solid particles. Thus, the *tortuosity* T for a porous medium may be defined by:

$$T = L_e/L, \quad T \geq 1. \tag{3.16}$$

This definition presents two problems: first, that of incorporating T appropriately in the expressions for ΔP; and secondly, that of quantifying T for fluid flow through particle beds.

The former problem involves a difficult choice. Either the tube length L can be simply replaced by TL in the above expressions for ΔP, a procedure which maintains the same fluid velocity in the 'equivalent tube' as in the particle bed but leads to different fluid-residence times; or alternatively the fluid residence times can be matched by allowing fluid velocities to differ by a factor of T. This latter procedure was proposed by Carman (1937) for packed beds, and supported more recently by Epstein (1989). The arguments presented below, however, relating particle drag to bed pressure loss, suggest the matching of fluid velocity as the more consistent alternative, and this we now adopt.

The viscous flow regime: revised tube-flow analogy

The substitutions to be made to the Hagen–Poiseuille equation, replacing those of eqn (3.9), now become:

$$L \rightarrow TL \qquad D \rightarrow D_e \qquad U \rightarrow U/\varepsilon, \tag{3.17}$$

leading, in place of eqn (3.10), to:

$$\Delta \mathrm{P} = 72 \cdot \frac{\mu_f L U}{d_p^2} \cdot \frac{(1-\varepsilon)^2}{\varepsilon^3} \cdot T. \tag{3.18}$$

The tortuosity relation

We are now faced with the problem of quantifying the tortuosity T itself. It is clear that fluid path lengths in concentrated particle beds will be significantly greater than the bed length L, but will approach L as the void fraction approaches unity. T must therefore be regarded as a function of ε. This becomes particularly important for flow through a fluidized bed, where ε varies with fluid flow rate.

The trend of T with void fraction (Figure 3.1) is captured by the simple relation:

$$T = 1/\varepsilon, \tag{3.19}$$

which converges to the correct, fully-expanded bed limit at $\varepsilon = 1$. This expression has been used to express tortuosity in the 'random pore model'

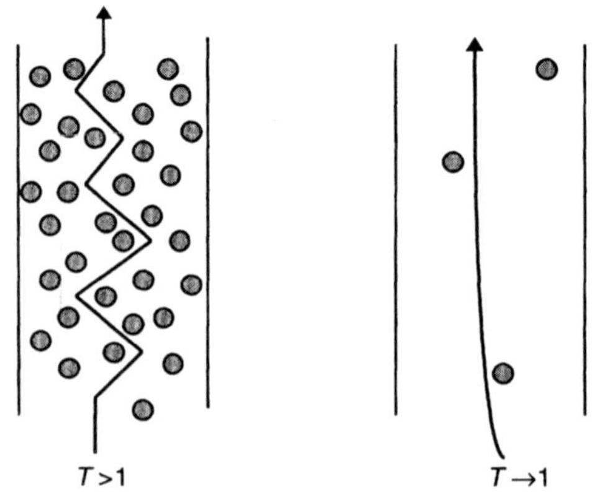

Figure 3.1 Tortuosity as a function of void fraction.

for diffusion in porous media (Wakao and Smith, 1962). It relates to a very simple probabilistic model for fluid flow through particle beds (Foscolo *et al.*, 1983), which regards the path of an element of fluid to consist of small steps, each of length δl. The fluid element takes a step forward if the way is clear, or laterally if the forward direction is blocked by a particle; the probabilities of these two alternatives may be regarded as being, respectively, ε and $(1-\varepsilon)$. On this basis, the probability tree of Figure 3.2 shows the various possible total path lengths for a fluid element that moves one step in the forward direction.

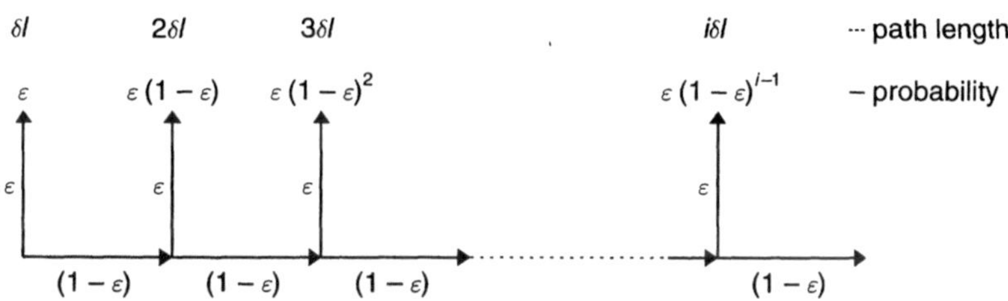

Figure 3.2 Probability tree for tortuosity.

From Figure 3.2 the tortuosity is seen to emerge as an infinite series, which sums conveniently to the expression of eqn (3.19):

$$T = \frac{\sum_i i\varepsilon(1-\varepsilon)^{i-1}\delta l}{\delta l} = \frac{1}{\varepsilon}. \tag{3.20}$$

A somewhat different expression for the tortuosity function has been suggested by Puncochar and Drahos (1993).

The unrecoverable pressure loss

Inserting the relation for T of eqn (3.19) in eqn (3.18), and making an empirical adjustment to the constant (from 72 to 60), produces an expression for ΔP that is in exact agreement with the Blake–Kozeny equation at the normal packed bed void fraction of 0.4; but which, beyond that point, reflects the above tortuosity considerations for expanded beds, $\varepsilon > 0.4$:

$$\Delta P = 60 \cdot \frac{\mu_f L U}{d_p^2} \cdot \frac{(1-\varepsilon)^2}{\varepsilon^4}. \tag{3.21}$$

The inertial flow regime: revised tube-flow analogy

The foregoing arguments regarding tortuosity call for the same substitutions, eqn (3.17) with T given by eqn (3.19), in the conventional inertial regime relation, eqn (3.13). This time, however, a further effect of bed expansion has to be taken into account.

The friction factor

The tube-flow equation for the inertial flow regime, eqn (3.11), is in terms of a friction factor f, which remains constant *for a given tube*. It varies considerably, however, with tube roughness, increasing markedly with the extent of tube wall imperfections that present obstructions normal to the direction of flow (Bird *et al.*, 1960). In the analogy relating tube flow to flow through beds of particles, it is the particles themselves that provide such obstructions. For an expanded bed, where the number of particles per unit length is less than for densely packed beds, the analogy would therefore require a reduction in effective tube roughness, and hence in the value for f. For a fluidized bed, an increasing fluid flow rate results in a continuously increasing void fraction, calling for a progressively decreasing f, approaching zero as ε approaches unity.

This trend, and the limiting values for f at $\varepsilon = 0.4$ and $\varepsilon = 1$, can be captured by simply setting f proportional to particle concentration:

$$\mathrm{f} \propto (1 - \varepsilon), \tag{3.22}$$

with the proportionality constant chosen to provide complete agreement with the inertial term of the Ergun equation at $\varepsilon = 0.4$.

The unrecoverable pressure loss

On including this friction factor dependency, eqn (3.22), along with the substitutions of eqns (3.17) and (3.19) in eqn (3.13), the unrecoverable pressure loss in the inertial regime becomes:

$$\Delta \mathrm{P} = 1.17 \cdot \frac{\rho_f L U^2}{d_p} \cdot \frac{(1 - \varepsilon)^2}{\varepsilon^4}. \tag{3.23}$$

A revised void fraction dependency for ΔP: the fully expanded bed limit

A convenient effect of the above changes to the conventional pressure loss expressions for the viscous and inertial flow regimes has been the unification of the dependencies on void fraction: these are now the same in eqns (3.21) and (3.23). There is, however, another factor to consider regarding the general applicability of these relations.

The analogy with tube flow, while providing a remarkably effective means for estimating ΔP in densely packed beds, becomes implausible for very dilute particle systems: here the more appropriate focus is the mechanism for flow past a single, unhindered particle, the limiting condition as $\varepsilon \to 1$. What are required, therefore, are expressions for ΔP based on the tube-flow analogy for concentrated beds, which agree with Ergun as $\varepsilon \to 0.4$, and at the same time approach the correct, unhindered particle limit as $\varepsilon \to 1$. As will now be demonstrated, eqns (3.21) and (3.23) fail to satisfy this final condition.

A satisfactory interpolation between the packed bed and fully expanded limits can be achieved by first considering a typical particle in a bed of many others. The unrecoverable pressure loss ΔP comes about as a result of energy dissipation in the bed, and is therefore directly related to particle drag f_d. We have expressions for particle drag for the unhindered case under low and high Reynolds number conditions: eqns (2.5)–(2.7). What are now required are counterpart expressions for a particle in a concentrated bed. These may be deduced from the above unrecoverable pressure loss expressions, eqns (3.21) and (3.23).

Relation of particle drag f_d to the unrecoverable pressure loss ΔP

Consider a control volume consisting of a uniform bed of particles of vertical length L and unit cross-sectional area. Energy dissipation in this control volume may be computed from two different viewpoints: first, by considering the difference in the total energy content of the fluid entering and leaving; secondly, by summing the dissipation brought about by individual particles. On equating these two quantities the relation of ΔP to f_d emerges as follows:

1. *Energy lost by the fluid: external viewpoint.* Energy dissipation is responsible for a pressure loss ΔP in the fluid passing through the

control volume. For a volumetric flux U this amounts to a rate of energy loss δE:

$$\delta E = U\Delta\text{P}. \tag{3.24}$$

2. *Energy loss within the control volume: internal viewpoint.* The velocity of fluid within the control volume is U/ε. The rate of energy dissipation associated with a single stationary particle, which experiences a drag force f_d, may therefore be taken to equal $f_d U/\varepsilon$. The number of such particles in the control volume is $6(1-\varepsilon)L/\pi d_p^3$, so that the total rate of energy loss is given by:

$$\delta E = \frac{6(1-\varepsilon)L}{\pi d_p^3} \cdot \frac{U f_d}{\varepsilon}. \tag{3.25}$$

Equating these two expressions for δE, eqns (3.24) and (3.25), provides a link between particle drag and unrecoverable pressure loss:

$$f_d = \frac{\pi d_p^3 \varepsilon}{6L(1-\varepsilon)} \cdot \Delta\text{P}. \tag{3.26}$$

This useful relation enables us to switch focus between energy dissipation associated with a single particle in a bed, and its effect on the bed as whole.

Viscous flow conditions

Applying eqn (3.26) to the derived pressure loss expression for viscous flow in particle beds, eqn (3.21), yields:

$$f_d = 10\pi d_p \mu_f U \cdot \frac{(1-\varepsilon)}{\varepsilon^3} = 3\pi d_p \mu_f U \cdot \frac{3.33(1-\varepsilon)}{\varepsilon^3}. \tag{3.27}$$

The drag force f_d on a particle in a bed thus emerges as the product of the unhindered expression, $3\pi d_p \mu_f U$, and a 'voidage function', $3.33(1-\varepsilon)/\varepsilon^3$.

Note that for a normal packed bed, $\varepsilon = 0.4$, the voidage function assumes a value in excess of 30; in the limit, as $\varepsilon \rightarrow 1$, it approaches zero. By simply adding 1 to the general voidage function expression, the correct, unhindered limit is approached without significantly affecting its value in a concentrated bed:

$$f_d = 3\pi d_p \mu_f U \cdot \left(\frac{3.33(1-\varepsilon)}{\varepsilon^3} + 1\right). \tag{3.28}$$

Inertial flow conditions

Applying the identical procedure for the inertial flow regime pressure-loss expression, eqn (3.23), also yields particle drag as the product of its unhindered value and a voidage function; once again, the addition of 1 to the derived voidage function, which ensures the correct unhindered limit, has negligible effect for void fractions corresponding to a concentrated bed.

$$f_d = 0.055\pi\rho_f d_p^2 U^2 \cdot \left(\frac{3.55(1-\varepsilon)}{\varepsilon^3} + 1\right). \tag{3.29}$$

Equations (3.28) and (3.29) provide interpolations of the drag force on a particle in a bed of particles, which are applicable over the entire range of achievable void fraction, up to the unhindered limit. The voidage functions (the bracketed expressions in these two equations) are numerically very similar, as is clear from Figure 3.3, which compares them over the full operating range. Also shown in this figure is the function $\varepsilon^{-3.8}$, which is likewise very similar numerically; from a practical point of view, these three forms may be regarded as interchangeable.

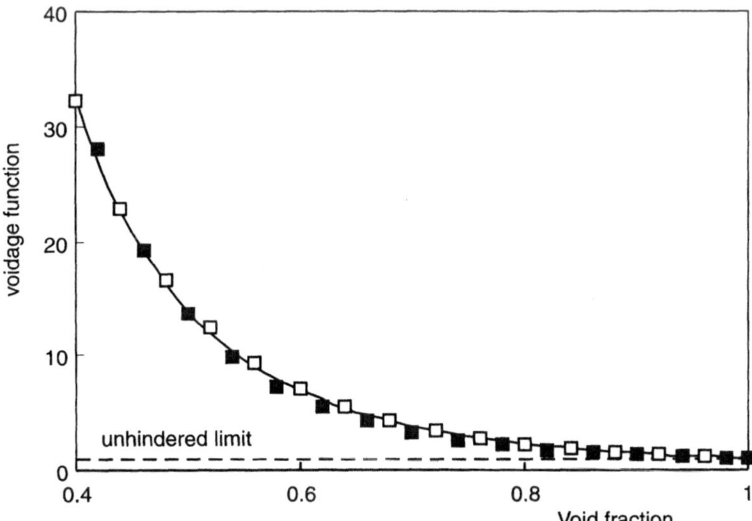

Figure 3.3 'Voidage functions' for drag on a particle in a particle bed: Continuous curve, the common adopted form, $\varepsilon^{-3.8}$; open squares, the viscous regime form, $3.33(1-\varepsilon)/\varepsilon^3 + 1$; solid squares, the inertial regime form, $3.55(1-\varepsilon)/\varepsilon^3 + 1$.

Particle drag in a uniform bed

We are therefore led to adopt the more compact form for the voidage function, leading to the following relations for the drag force on a particle in a particle bed:

$$f_d = 3\pi d_p \mu_f U \cdot \varepsilon^{-3.8}, \quad \textit{Viscous regime} \tag{3.30}$$

$$f_d = 0.055\pi \rho_f d_p^2 U^2 \cdot \varepsilon^{-3.8}, \quad \textit{Inertial regime} \tag{3.31}$$

which both converge to the unhindered particle limits, eqns (2.5)–(2.7), as $\varepsilon \rightarrow 1$.

Unrecoverable pressure loss in a particle bed

We may now apply the relation between particle drag and unrecoverable pressure loss, eqn (3.26), to eqns (3.30) and (3.31) to yield the pressure loss equations, applicable over the full expansion range, $1 \geq \varepsilon \geq 0.4$:

$$\Delta \mathrm{P} = 18 \cdot \frac{\mu_f L U}{d_p^2} \cdot (1 - \varepsilon)\varepsilon^{-4.8}, \quad \textit{Viscous regime} \tag{3.32}$$

$$\Delta \mathrm{P} = 0.33 \cdot \frac{\rho_f L U^2}{d_p} \cdot (1 - \varepsilon)\varepsilon^{-4.8}, \quad \textit{Inertial regime} \tag{3.33}$$

so that the revised Ergun equation becomes:

$$\Delta \mathrm{P} = \left(\frac{18}{Re_p} + 0.33\right) \cdot \frac{\rho_f U^2 L}{d_p} \cdot (1 - \varepsilon)\varepsilon^{-4.8}. \tag{3.34}$$

This relation agrees with the usual Ergun form, eqn (3.15), for normal packed beds with $\varepsilon \approx 0.4$. For expanded beds eqn (3.34) deviates progressively from the Ergun equation, reflecting the decreasing tortuosity and inertial regime friction factor with increasing ε, and the approach to single particle suspension as ε approaches unity.

We have said nothing in the preceding discussion about the physical significance of high void fraction beds. How can such arrangements be achieved in practice? One obvious possibility is for the bed to be fluidized homogeneously, and that, it must be admitted, has been the major incentive for developing eqns (3.30)–(3.34). However, it should be emphasized that no assumptions whatsoever concerning the fluidized state have gone into uncovering these relations. Their applicability to fluidized beds will be demonstrated in Chapter 4. For now, we round off the discussion by comparing predictions of eqn (3.34) with reported measurements in particle beds that have been expanded to high void fraction by less conventional means.

Experiments on expanded fixed beds of spheres

The earliest attempt at measuring ΔP in expanded beds involved inserting thin rods, threaded with 5 mm spherical beads, into vertical tubes (Happel and Epstein, 1954; Figure 3.4). The beads and rods were carefully spaced and arranged so as to produce beds containing uniform, cubical arrays of spheres, with void fractions ranging from 0.69 to 0.94. Measurements of ΔP in glycerol solutions flowing through four such beds were reported as fitted functions of the particle Reynolds number. (A somewhat similar arrangement was later adopted by Rowe (1961) for the study of fluid interaction with a single particle placed within the particle matrix.)

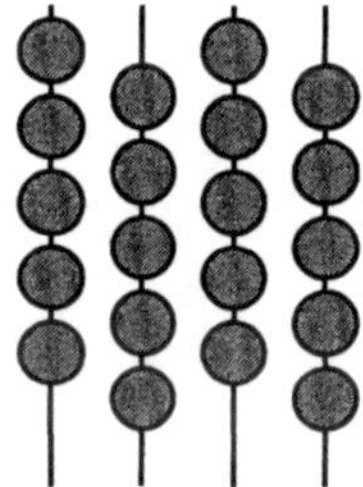

Figure 3.4 The expanded fixed beds of Happel and Epstein (1954).

A second investigation (Wentz and Thodos, 1963a, 1963b), carried out at much higher particle Reynolds numbers, involved air, a standard wind tunnel and cubical arrays of 31 mm spheres joined together by means of fine wires. Five such assemblies were constructed, void fractions ranging from 0.48 to 0.88.

Finally, an ingenious technique was employed to produce *randomly packed*, high void fraction beds (Rumpf and Gupte, 1971; Figure 3.5). This involved first packing a mixture of polystyrene spheres and sugar particles in a tube, which was then flushed with carbon tetrachloride; this attacked the polystyrene surfaces, making them sticky and thus causing the spheres to weld together at contact points; finally, the sugar particles, which served solely to increase the average space separating the spheres, were dissolved away with water. The final result of this operation was a rigid, randomly orientated structure having a void fraction ε in the range 0.41–0.64. Pressure drop measurements were reported in seven such units for both gas and liquid flows.

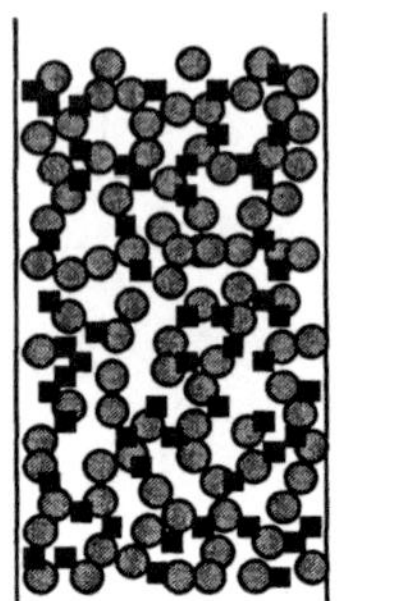
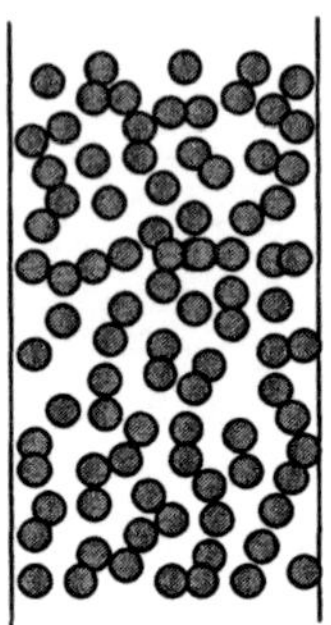

Figure 3.5 A random-packed expanded bed (Rumpf and Gupte, 1971): sugar particles, which separate the spheres (left) are dissolved away to yield an expanded structure (right).

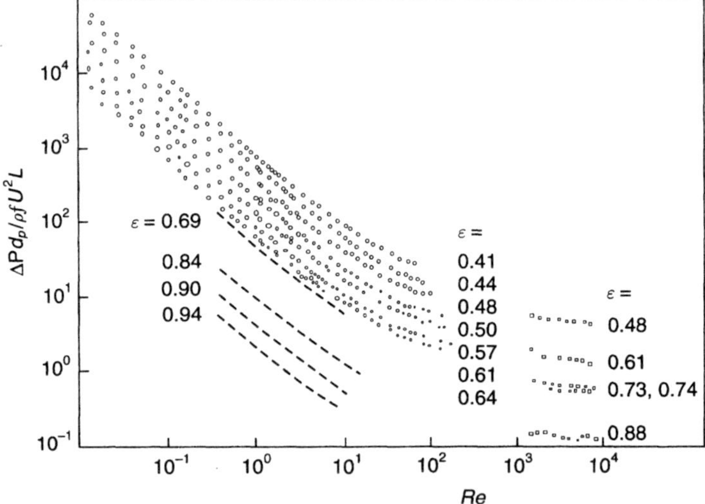

Figure 3.6 Fluid pressure loss in expanded fixed beds: experimental measurements. Broken lines, data of Happel and Epstein (1954); squares, data of Wentz and Thodos (1963a, 1963b); circles, data of Rumpf and Gupte (1971).

The results of these three investigations are shown in Figure 3.6, the first as broken lines representing the published correlations, the other two as raw data points. The range covered is enormous: six orders of magnitude in both dimensionless unrecoverable pressure loss and particle Reynolds number – from well inside the viscous to deep into the inertial flow regimes.

Figure 3.7 shows the correlation of these data in terms of the general pressure drop relation, eqn (3.34). It will be seen that the effect of the

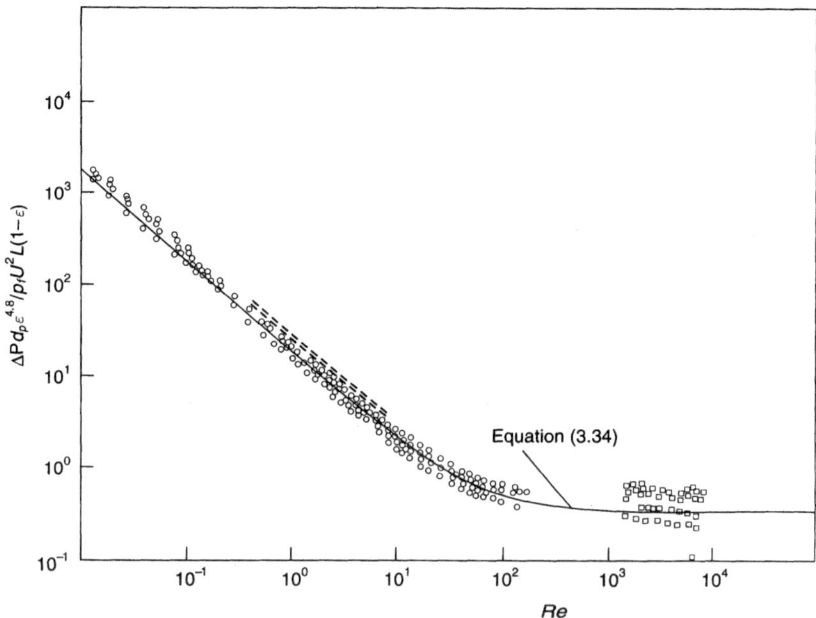

Figure 3.7 Fluid pressure loss in expanded fixed beds: comparison of all measurements reported in Figure 3.6, with the predictions of eqn (3.34) (continuous curve); data symbols as for Figure 3.6.

revised void fraction dependence is to draw all the results close to the predicted expression. Even the small remaining spread in the very high Re_p results can be tentatively attributed to the effect of the connecting wires, which were not corrected for in these wind tunnel experiments (Gibilaro *et al.*, 1985).

The comparisons shown in Figure 3.6 are encouraging, supporting as they do the predictive ability of the derived, unrecoverable pressure loss expression, eqn (3.37), for fluid flow through beds of spheres over the full expansion range encountered with fluidized systems.

References

Bird, R.B., Stewart, W.E. and Lightfoot, E.N. (1960). *Transport Phenomena*. Wiley.

Carman, P.C. (1937). Fluid flow through a granular bed. *Trans. Inst. Chem. Engrs.*, **15**, 150.

Epstein, N. (1989). On tortuosity and the tortuosity factor in flow and diffusion through porous media. *Chem. Eng. Sci.*, **44**, 777.

Ergun, S. and Orning, A.A. (1949). Fluid flow through randomly packed columns and fluidized beds. *Ind. Eng. Chem.*, **41(6)**, 1179.

Foscolo, P.U., Gibilaro, L.G. and Waldram, S.P. (1983). A unified model for particulate expansion of fluidized beds and flow in porous media. *Chem. Eng. Sci.*, **38**, 1251.

Gibilaro, L.G., Di Felice, R., Waldram, S.P. and Foscolo, P.U. (1985). Generalised friction factor and drag coefficient correlations for fluid–particle interactions. *Chem. Eng. Sci.*, **40**, 1817.

Happel, J. and Epstein, N. (1954). Viscous flow in multiparticle systems: cubical assemblages of uniform spheres. *Ind. Eng. Chem.*, **46**, 1187.

Puncochar, M. and Drahos, J. (1993). The tortuosity concept in fixed and fluidized beds. *Chem. Eng. Sci.*, **48**, 2173.

Rowe, P.N. (1961). Drag forces in a hydraulic model of a fluidized bed: Part II. *Trans. Inst. Chem. Engrs.*, **39**, 175.

Rumpf, H. and Gupte, A.R. (1971). Influence of porosity and particle size distribution in resistance law of porous flow. *Chemie-Ing. Technik.*, **43**, 367.

Wakao, N. and Smith, J.M. (1962). Diffusion in catalyst pellets. *Chem. Eng. Sci.*, **17**, 825.

Wentz, C.A. Jr and Thodos, G. (1963a). Pressure drops in the flow of gases through packed and distended beds of spherical particles. *AIChE J.*, **9**, 81.

Wentz, C.A. Jr and Thodos, G. (1963b). Total and form drag friction factors for the turbulent flow of air through packed and distended beds of spheres. *AIChE J.*, **9**, 358.

4

Homogeneous fluidization

The unrecoverable pressure loss for fluidization

Expressions were derived in the previous chapter for the unrecoverable pressure loss ΔP in a fluid flowing through a bed of particles. These were shown to apply to beds expanded by various mechanical means to void fractions normally encountered only in fluidized systems. In this chapter we make use of these relations in an analysis of the equilibrium state of homogeneous fluidization.

The steady-state balance of forces for a fluidized suspension

Consider a control volume of unit cross-sectional area and height L in a fluidized bed. The only surface forces we need consider in the axial direction are provided at the two horizontal boundary cross-sections by the fluid pressure; the net effect of these surface forces is to support the total weight of particles and fluid in the control volume:

$$\begin{aligned}\Delta p &= p(z) - p(z+L) \\ &= \big(\varepsilon\rho_f + (1-\varepsilon)\rho_p\big)Lg. \qquad (4.1)\end{aligned}$$

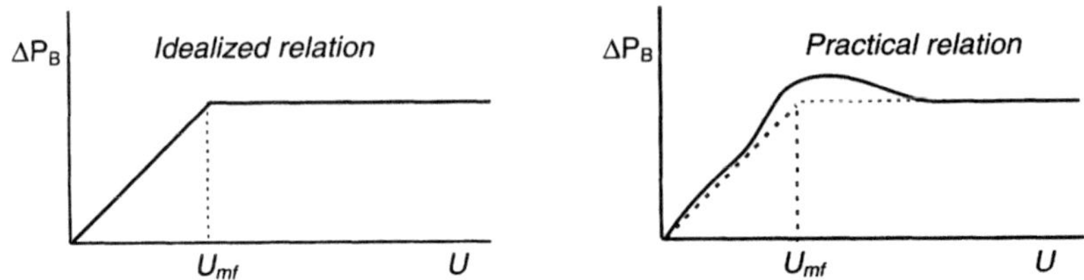

Figure 4.1 Unrecoverable pressure loss in a fluidized bed.

The unrecoverable pressure loss, an indelible consequence of maintaining the particles in suspension, is thus:

$$\Delta P = \Delta p - \rho_f Lg = (\rho_p - \rho_f)(1 - \varepsilon)Lg. \tag{4.2}$$

An important property of fluidized beds follows immediately from this simple relation. If we apply it to the whole bed, of height L_B, rather than just a fixed slice of height L, then the product $(1 - \varepsilon)L_B$ represents the total volume V_B of particles per unit cross-section, which remains unchanged as the bed expands: as the fluid flux is increased, L_B increases and $(1 - \varepsilon)$ decreases so as to maintain their product at a constant value. Thus the unrecoverable pressure loss ΔP_B for the whole bed becomes:

$$\Delta P_B = (\rho_p - \rho_f)V_B g = a\ constant. \tag{4.3}$$

This well-known relation is illustrated in Figure 4.1. In practice, the transition between the fixed and fluidized states involves some particle rearrangement, with the breakdown of bridging structures, which are inherent in the initial packing and subsequent defluidization operations; rather than an abrupt change in slope at the minimum fluidization velocity U_{mf}, a more gradual approach to the constant ΔP_B is observed in practice, often with some overshoot in the transition region.

Steady-state expansion of fluidized beds

The expansion characteristics of homogeneously fluidized beds have been the subject of far more empirical study than theoretical analysis. This could be due to the uncomplicated nature of the experimental procedure, which involves simply the measurement of steady-state bed height L_B as a function of volumetric fluid flux U. The results are usually presented as the relation of U with void fraction ε, which, unlike L_B, is independent of the quantity of particles present. The constant particle volume relation,

$V_B = L_B(1 - \varepsilon)$, enables ε to be calculated from L_B from the initial values of these variables in the packed bed. Before applying the ΔP relations (derived in the previous chapter) to the analysis of fluidized bed expansion, a brief account of the salient experimental findings will be given.

Empirical results

It has been widely verified that a plot of U against ε on logarithmic co-ordinates approximates closely to a straight line over the full range of bed expansion, regardless of the flow regime. Small deviations from this behaviour, reported for very high void fractions, $\varepsilon > 0.95$ (Garside and Al-Dibouni, 1977; Rapagnà *et al.*, 1989), need not concern us at this stage. The observations may therefore be described by:

$$U = u_t \varepsilon^n. \tag{4.4}$$

The relation shown in eqn (4.4) appears to have been first observed by Lewis *et al.* (1949), but is now universally known as the *Richardson–Zaki* equation after the authors of an extensive experimental investigation into its applicability (Richardson and Zaki, 1954a, 1954b; Figure 4.2).

The parameter n correlates with the terminal particle Reynolds number Re_t: it acquires constant values in both the creeping flow and inertial flow regimes ($n \approx 4.8$ and 2.4 respectively), changing progressively with Re_t in the intermediate regime between these limiting values. The following convenient relation (Khan and Richardson, 1989) enables n to be

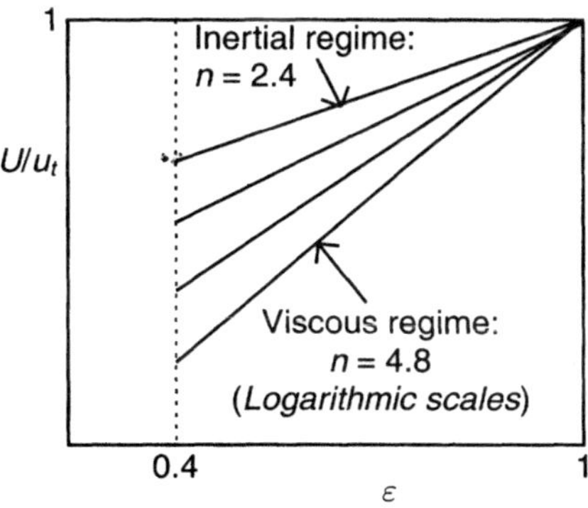

Figure 4.2 Steady-state expansion characteristics for homogeneous fluidization: the Richardson–Zaki relation.

evaluated from the Archimedes number Ar, which is in terms of solely the fluid and particle properties, $Ar = gd_p^3\rho_f(\rho_p - \rho_f)/\mu_f^2$, rather than Re_t.

$$\frac{4.8 - n}{n - 2.4} = 0.043Ar^{0.57}. \tag{4.5}$$

It is clear that for viscous conditions, small Ar, this converges to $n = 4.8$, and for inertial conditions, large Ar, to $n = 2.4$.

The viscous flow regime

The expansion characteristics may be obtained by equating the ΔP relation for viscous flow through beds of spheres, eqn (3.32), to that required for supporting a fluidized suspension, eqn (4.2). This yields:

$$U = \frac{(\rho_p - \rho_f)gd_p^2}{18\mu_f} \cdot \varepsilon^{4.8} = u_t\varepsilon^{4.8}. \tag{4.6}$$

This result is in complete agreement with the empirical Richardson–Zaki relation for the viscous flow regime: eqn (4.4), $n = 4.8$. This is a satisfying conclusion, as eqn (3.32) was formulated solely in terms of fluid flow through beds of particles, quite independently of any relation to the fluidized state. The limit as $\varepsilon \rightarrow 1$ yields *Stokes law*, eqn (2.12), the analytical form for u_t; this, however, is unsurprising as the unhindered-particle limiting condition was specifically imposed in the derivation of eqn (3.32).

The inertial flow regime

The counterpart procedure for the inertial flow regime, equating the ΔP expression of eqn (3.33) to eqn (4.2), yields;

$$U = \sqrt{3.03gd_p(\rho_p - \rho_f)/\rho_f} \cdot \varepsilon^{2.4} = u_t\varepsilon^{2.4}. \tag{4.7}$$

Once again we have arrived, quite independently, at a result for the expansion characteristics of homogeneous fluidized beds which is in compete agreement with the Richardson–Zaki relation, this time for inertial flow conditions: eqn (4.4), $n = 2.4$. The unhindered-particle limit, $\varepsilon \rightarrow 1$, yields, as it must, the inertial regime relation of eqn (2.13) for u_t.

All flow regimes

A general expression for the expansion characteristics will now be derived on the basis of the following constitutive relation for the unrecoverable pressure loss over the bed as a whole:

$$\Delta \mathrm{P_B} \propto U^a \varepsilon^b. \tag{4.8}$$

This form corresponds to the viscous and inertial relations, eqns (3.32) and (3.33), applied to the entire bed ($L = L_B$) for which the product $(1 - \varepsilon)L_B$ becomes a constant.

We have seen that changes in the fluid flux give rise to changes in void fraction that maintain $\Delta \mathrm{P_B}$ constant. Therefore:

$$\mathrm{d}(\Delta \mathrm{P_B}) = \frac{\partial \Delta \mathrm{P_B}}{\partial U} \cdot \mathrm{d}U + \frac{\partial \Delta \mathrm{P_B}}{\partial \varepsilon} \cdot \mathrm{d}\varepsilon = 0, \tag{4.9}$$

$$\frac{\mathrm{d}U}{\mathrm{d}\varepsilon} = -\frac{\partial \Delta \mathrm{P_B}}{\partial \varepsilon} \Big/ \frac{\partial \Delta \mathrm{P_B}}{\partial U}. \tag{4.10}$$

On evaluating the partial derivatives in eqn (4.10) from the constitutive expression for $\Delta \mathrm{P_B}$, eqn (4.8), we arrive at the differential equation relating void fraction to fluid flux:

$$\frac{\mathrm{d}U}{\mathrm{d}\varepsilon} = -\frac{\mathrm{b}U}{\mathrm{a}\varepsilon}. \tag{4.11}$$

Solving eqn (4.11) with boundary condition $\varepsilon = 1$, $U = u_t$, yields a general expression for the expansion characteristics:

$$U = u_t \varepsilon^{-\mathrm{b/a}}. \tag{4.12}$$

This form is identical to the Richardson–Zaki equation, eqn (4.4). It therefore relates the parameter n in that empirical relation to the ratio of the void fraction and fluid flux exponents in the expression for unrecoverable pressure loss, eqn (4.8): $n = -\mathrm{b/a}$. Note that under both viscous and inertial flow conditions ($\mathrm{a} = 1$, $n = 4.8$ and $\mathrm{a} = 2$, $n = 2.4$, respectively), the void fraction exponent b assumes the value of -4.8. This unexpected coincidence will now be put to effective use.

Two working hypotheses. The above interpretation of the empirical parameter n, together with the evidence for effectively identical void fraction dependencies in the unrecoverable pressure loss relations for the viscous and inertial flow regimes, suggests the following generalization for fluidization, applicable to all flow regimes:

$$\Delta P_B \propto U^{4.8/n} \varepsilon^{-4.8}, \tag{4.13}$$

$$\Delta P \propto U^{4.8/n} \varepsilon^{-4.8} (1 - \varepsilon). \tag{4.14}$$

The fluid flux exponent, $4.8/n$, in eqns (4.13) and (4.14) converges to the correct limits of 1 and 2 for the viscous and inertial flow regimes, for which n has values of 4.8 and 2.4 respectively. In the intermediate regime it serves to provide a convenient, if approximate, interpolation between these two extremes. The value of -4.8 for the void fraction exponent also represents an approximation for intermediate flow conditions: values as different as -4.2 have been reported for Reynolds numbers of around 50 where the maximum deviation appears to occur (Khan and Richardson, 1990; Di Felice, 1994). These reservations are of secondary relevance, however, pointing only to the possibility of minor quantitative inaccuracies in predictions arising from analyses in which the relations of eqns (4.13) and (4.14) are applied.

The primary forces acting on a fluidized particle

In Chapter 2, the primary forces acting on a single particle in a flowing fluid were quantified and applied to the determination of the terminal settling velocity u_t. Expressions were derived for u_t in terms of the basic fluid and particle properties (ρ_f, μ_f, ρ_p, d_p). In this section we derive the counterpart relations for a particle in a fluidized bed (Foscolo *et al.*, 1983; Foscolo and Gibilaro, 1984). This represents an important step in the analysis of the fluidized state, in which large numbers of particles are held simultaneously in suspension. Just as for single particle suspension, the primary forces acting on a fluidized particle may be identified as the effects of gravity, buoyancy (the net result of the mean fluid pressure gradient to which the particle is subjected), and drag.

The buoyancy force

Consider the particle of arbitrary shape shown in Figure 4.3. Its total projected area in the horizontal plane is A; a small element dA of this area corresponds to the top and bottom of an element of particle volume of

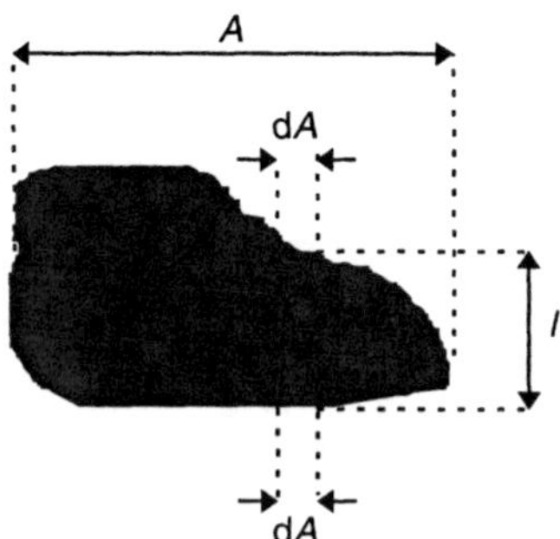

Figure 4.3 Particle buoyancy in a fluidized suspension.

vertical height l. The particle is situated in a fluidized bed; it may be regarded either as a typical component of the fluidized inventory, or simply as an extraneous object supported in the bed in some manner.

The difference in pressure Δp between the bottom and top face of this volume element gives rise to a net force df_b acting on the two horizontal projected area elements:

$$df_b = \Delta p dA. \tag{4.15}$$

For situations in which the pressure gradient in the fluid, dp/dz, may be regarded as constant, Δp is given by:

$$\Delta p = -l\frac{dp}{dz}. \tag{4.16}$$

Inserting this relation into eqn (4.15) and integrating over the projected area of the particle yields:

$$f_b = -\frac{dp}{dz}\int_A l dA = -V_p\frac{dp}{dz}, \tag{4.17}$$

where V_p is the particle volume.

The principle of Archimedes was conceived for the situation in which the pressure gradient to which a submerged body is subjected results simply from the action of a gravitational field on a static fluid, $dp/dz = -\rho_f g$, leading to: $f_b = V_p\rho_f g = -$ 'weight of displaced fluid'. Eqn (4.17) represents a more general statement of this well known result: a linear pressure field, *regardless of its origin*, imparts a force on a body placed within it that is equal in magnitude to the product of the pressure gradient and the volume of the submerged body.

The fluid pressure gradient in a fluidized bed under equilibrium conditions follows from eqn (4.1):

$$\frac{\mathrm{d}p}{\mathrm{d}z} = -(\varepsilon\rho_f + (1-\varepsilon)\rho_p)g. \tag{4.18}$$

Inserting this relation into eqn (4.17) delivers an expression for buoyancy, which for a sphere becomes:

$$f_b = \frac{\pi d_p^3}{6}(\varepsilon\rho_f + (1-\varepsilon)\rho_p)g. \tag{4.19}$$

The expression in parentheses in eqn (4.19) represents the mean density of the fluidized suspension ρ_s: $\rho_s = (\varepsilon\rho_f + (1-\varepsilon)\rho_p)$. What this result demonstrates is that in applying Archimedes principle to a body in a steady-state fluidized bed it is suspension, rather than fluid, that the body may be thought to 'displace'. For an object immersed in a pure fluid, $\varepsilon = 1$, eqn (4.19) reduces to the familiar form.

The more general buoyancy expression, eqn (4.17), is applied to the case of *non-equilibrium* fluidized suspensions in Chapter 11.

The effective weight of a fluidized particle

For single particle suspension, it was found convenient to define the *effective weight* w_e of a particle as the net effect of gravity and buoyancy: eqn (2.9). On applying the same definition to a fluidized sphere in equilibrium, for which,

$$f_g = -\frac{\pi d_p^3}{6}\rho_p g, \qquad f_b = \frac{\pi d_p^3}{6}(\varepsilon\rho_f + (1-\varepsilon)\rho_p)g, \tag{4.20}$$

we obtain:

$$w_e = f_g + f_b = -\frac{\pi d_p^3}{6}(\rho_p - \rho_f)g\varepsilon. \tag{4.21}$$

This pleasingly simple relation shows the effective weight of an 'average' fluidized particle under equilibrium conditions to be simply proportional to the void fraction.

The drag force

In Chapter 3, expressions for the drag force f_d on a particle in a bed of particles were obtained from the relations for unrecoverable pressure loss ΔP. These apply quite generally, regardless of how the particles are supported. They will now be applied to particles in a fluidized bed.

The viscous flow regime

Under viscous flow conditions, particle drag was found to be given by eqn (3.30):

$$f_d = 3\pi d_p \mu_f U \cdot \varepsilon^{-3.8}, \tag{4.22}$$

which may be written:

$$f_d = 3\pi d_p \mu_f u_t \cdot \left(\frac{U}{u_t}\right) \cdot \varepsilon^{-3.8}. \tag{4.23}$$

The first part of this expression, $3\pi d_p \mu_f u_t$, represents the drag force required to just suspend a single, unhindered particle; it therefore equates to the particle's effective weight, $\pi d_p^3 (\rho_p - \rho_f) g/6$, leading to:

$$f_d = \frac{\pi d_p^3}{6} \cdot (\rho_p - \rho_f) g \cdot \left(\frac{U}{u_t}\right) \cdot \varepsilon^{-3.8}. \tag{4.24}$$

This form is readily amenable to generalization.

All flow regimes

Adopting the general form for ΔP proposed in the previous section, eqn (4.14), and invoking once again the relation linking particle drag to unrecoverable pressure loss, eqn (3.26), leads to the generalization of eqn (4.24) for all flow regimes:

$$f_d = \frac{\pi d_p^3}{6} \cdot (\rho_p - \rho_f) g \cdot \left(\frac{U}{u_t}\right)^{\frac{4.8}{n}} \cdot \varepsilon^{-3.8}. \tag{4.25}$$

The consistency of this expression may be confirmed for conditions of equilibrium by equating it to the particle effective weight w_e, eqn (4.21): this yields $U = u_t \varepsilon^n$, the ubiquitous Richardson–Zaki relation.

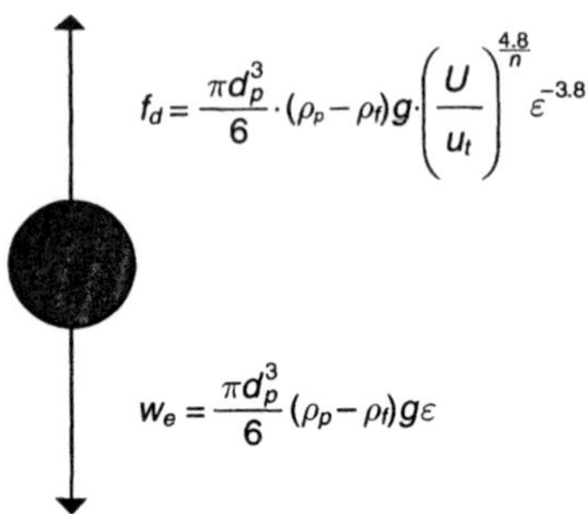

Figure 4.4 The primary forces acting on a fluidized particle.

We have thus obtained explicit expressions, in terms of the known parameters, for the primary forces that act on a fluidized particle (Figure 4.4). These were derived on the basis of steady-state, equilibrium assumptions and validated in a variety of ways for the equilibrium state. When we come to apply them in later chapters to the analysis of non-equilibrium behaviour, it will be found that other considerations, also of a quantifiable nature, need to be taken into account.

References

Di Felice, R. (1994). The voidage function for fluid–particle interaction systems. *Int. J. Multiphase Flow*, **20**, 153.

Foscolo, P.U. and Gibilaro, L.G. (1984). A fully predictive criterion for the transition between particulate and aggregate fluidization. *Chem. Eng. Sci.*, **39**, 1667.

Foscolo, P.U., Gibilaro, L.G. and Waldram, S.P. (1983). A unified model for particulate expansion of fluidized beds and flow in porous media. *Chem. Eng. Sci.*, **38**, 1251.

Garside, J. and Al-Dibouni, M.R. (1977). Velocity-voidage relationships for fluidization and sedimentation in solid–liquid systems. *Chem. Eng. Sci.*, **16**, 206.

Khan, A.R. and Richardson, J.F. (1989). Fluid–particle interactions and flow characteristics of fluidized beds and settling suspensions of spherical particles. *Chem. Eng. Comm.*, **78**, 111.

Khan, A.R. and Richardson, J.F. (1990). Pressure gradient and friction factor for sedimentation and fluidization of uniform spheres in liquids. *Chem. Eng. Sci.*, **45**, 255.

Lewis, W.K., Gilliland, E.R. and Bauer, W.C. (1949). Characteristics of fluidized particles. *Ind. Eng. Chem.*, **41**, 1104.

Rapagnà, S., Di Felice, R., Gibilaro, L.G. and Foscolo, P.U. (1989). Steady-state expansion characteristics of beds of monosize spheres fluidized by liquids. *Chem. Eng. Comm.*, **79**, 131.

Richardson, J.F. and Zaki, W.N. (1954a). Sedimentation and fluidization. *Trans. Inst. Chem. Eng.*, **32**, 35.

Richardson, J.F. and Zaki, W.N. (1954b). The sedimentation of a suspension of uniform spheres under conditions of viscous flow. *Chem. Eng. Sci.*, **3**, 65.

5

A kinematic description of unsteady-state behaviour

The response of homogeneously fluidized beds to sudden changes in fluid flux

The responses described in this chapter are more usually associated with liquid-fluidized beds, which are more likely to fluidize homogeneously than gas beds. We start by considering the effect of relatively large step changes in the fluid flux to a bed initially in equilibrium, describing an idealized, qualitative mechanism for the transition to a new equilibrium state (Gibilaro *et al.*, 1984). The mechanism is somewhat different for decreases in fluid flux than for increases. First we will consider the former case, the contracting bed, which is the more straightforward, and then go on to consider the expanding bed, which introduces the concept of *interface stability*, a key factor in the formation and subsequent behaviour of fluidized suspensions.

The contracting bed

Consider a homogeneously fluidized bed of void fraction ε_1 in equilibrium with a fluid flux of U_1. At time zero the flux is suddenly switched to a lower value U_2, causing the bed to contract, eventually attaining a new equilibrium at void fraction ε_2.

The immediate effect of the drop in fluid flux is to bring about a sudden reduction in the fluid–particle interaction force on all the particles in the bed. These therefore experience a net force, causing them to accelerate downwards together, without any change in the void fraction ε_1; this particle acceleration results in a progressive increase in the relative fluid–particle velocity, causing, in turn, a progressive increase in particle drag. The process continues until such time as the relative velocity, and hence the interaction force, returns to the equilibrium value experienced by the particles prior to the change in fluid flux. From this point on the particles continue their downward motion at constant velocity, in equilibrium and still at void fraction ε_1.

The behaviour just described is clearly not possible for particles at the bottom of the bed, in contact with the distributor: these cannot move downwards and so remain stationary, soon to be joined by others arriving from above. This gives rise to a growing zone of stationary particles at the bottom of the bed, which adjusts to the equilibrium void fraction corresponding to zero particle velocity and the new fluid-flux U_2; this is ε_2, the value eventually to be reached by the whole bed after the transient rearrangement period has been concluded.

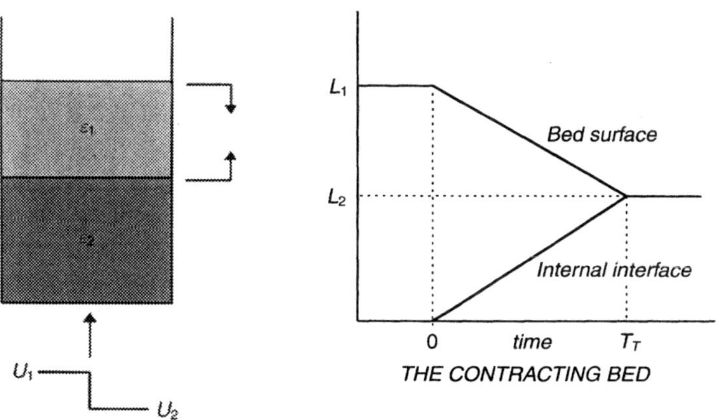

Figure 5.1 Idealized description of bed contraction.

The overall picture (following the initial period of particle acceleration, which usually amounts to a very small fraction of the total transient response time) is thus of two zones, both in fluid–particle equilibrium: an upper zone at void fraction ε_1, in which the particles are all falling at a constant velocity, and a lower zone of stationary particles at void fraction ε_2. There are thus two travelling interfaces: the falling surface of the bed, and the rising discontinuity, or *shock wave*, that separates the two zones (Figure 5.1). When these meet the whole bed will have attained the new equilibrium condition: U_2, ε_2.

The expanding bed

A sudden increase in fluid flux from U_1 to U_2 gives rise to a net upward force on all the particles; these therefore immediately start to accelerate upwards together, without change in void fraction ε_1, until the relative fluid–particle velocity and the interaction force drop back to their previous equilibrium levels; from this point on it would appear that the bed should continue its upward motion, piston-like at constant velocity, as was suggested in Chapter 1 on the basis of the analogy with single-particle suspension. The fact that this does not happen in practice (fluidized beds would never form if it did) has to do with the instability of the interface separating the bottom of the particle piston from the clear fluid below.

Interface stability

The particle piston, created as described above, possesses two interfaces with the fluid through which it travels. The stability of each can be determined on the basis of the following simple qualitative considerations.

The top interface. Imagine the top interface to be subjected to a small disturbance, which displaces a particle some way into the clear fluid above (Figure 5.2). The displaced particle immediately encounters a reduced fluid velocity, and a consequential reduction in drag; this results in a net downward force, which quickly returns the particle to its previous position. The top interface is therefore *stable*, a fact well confirmed by experiment.

The bottom interface. Now consider the counterpart situation at the bottom interface. Here a displaced particle also experiences a net downward force, a result of the reduction in fluid velocity which it

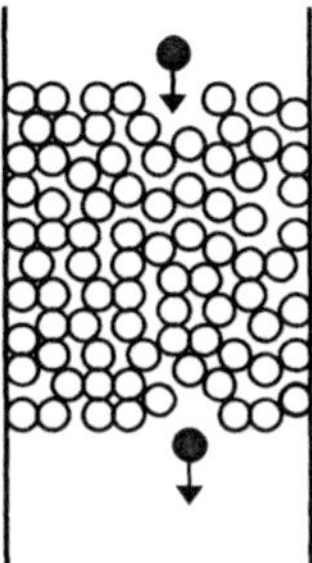

Figure 5.2 Stability tests on top and bottom interfaces.

encounters in the clear fluid below; but, far from having a stabilizing influence, the effect this time is to drive it down further from the interface, to be followed by particles from adjacent locations as they respond to the resulting increase in void fraction around them. The bottom interface is therefore *unstable*. Particles rain down from it continuously, giving rise to an upwards propagating erosion of the particle piston as it rises through the containing tube.

The above mechanism is set in motion very quickly, as soon as the particle piston starts to rise. The particles falling from it are stopped at the distributor where, for the same reason described for the contracting bed, they adjust to void fraction ε_2, in equilibrium with the new fluid flux U_2. The picture advanced by this somewhat idealized description, Figure 5.3, turns out to be very similar to that of the contracting bed: two equilibrium zones separated by an upwards propagating interface; the bottom zone

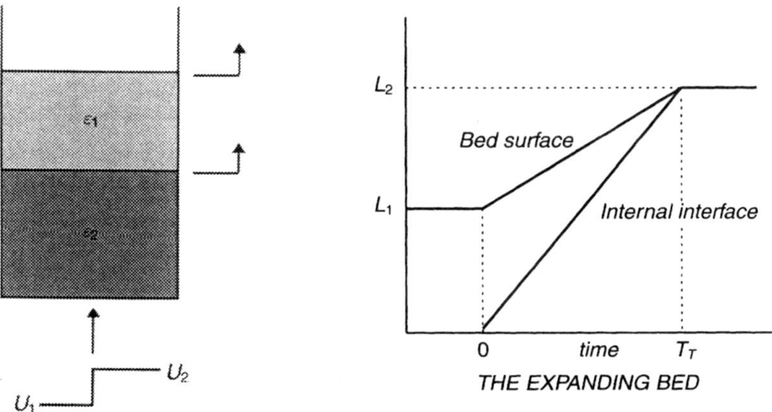

Figure 5.3 Idealized description of bed expansion.

consisting of stationary particles at the equilibrium condition eventually to be satisfied by the entire bed; the top zone consisting of particles at void fraction ε_1, travelling (upwards this time) at the constant velocity that maintains them at the equilibrium condition that existed before the fluid flux change. The lower interface, separating the two zones, travels faster than the bed surface, catching up with it at the completion of the rearrangement process.

The transient response of the bed surface

We are now in a position to quantify the above descriptions. Figure 5.4 relates to both contracting and expanding beds of unit cross-sectional area. The total bed height is L_B and the height of the interface separating the two zones is L_I. Both L_B and L_I are functions of time. The fluid flux in the bottom zone is U_2, and in the top zone is the yet to be determined U_A.

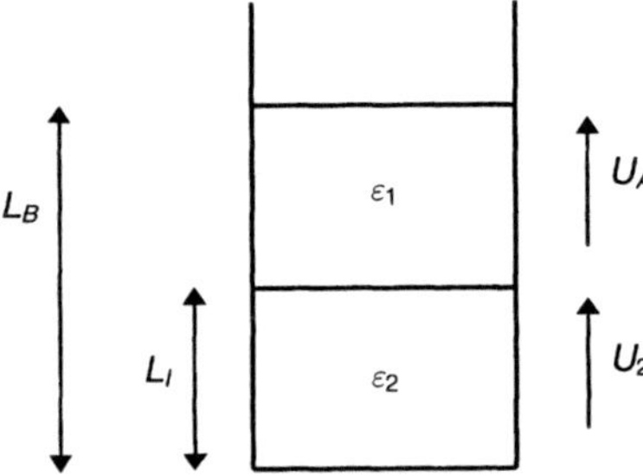

Figure 5.4 Transient response of homogeneously fluidized beds.

Mass balance for fluid in the bottom zone

The bottom zone is growing at a rate $\mathrm{d}L_I/\mathrm{d}t$. As it grows, the fluid content of its additional volume changes from ε_1 to ε_2. The rate of accumulation of fluid in this growing zone is therefore $(\varepsilon_2 - \varepsilon_1)\,\mathrm{d}L_I/\mathrm{d}t$, and the mass balance for fluid in this zone is given by:

$$U_2 - U_A = (\varepsilon_2 - \varepsilon_1)\frac{\mathrm{d}L_I}{\mathrm{d}t}. \tag{5.1}$$

U_A is the fluid flux that maintains equilibrium conditions in the top zone during the transient response period. As the void fraction ε_1 remains unchanged from its initial value, so must the relative fluid–particle velocity. The velocities of the fluid and particles in the upper zone are U_A/ε_1 and $\mathrm{d}L_B/\mathrm{d}t$ respectively, so that equating relative fluid–particle velocities before and during the transient response period yields:

$$\frac{U_A}{\varepsilon_1} - \frac{\mathrm{d}L_B}{\mathrm{d}t} = \frac{U_1}{\varepsilon_1}. \tag{5.2}$$

On combining eqns (5.1) and (5.2) to eliminate U_A, we obtain an equation linking the two interface velocities:

$$\varepsilon_1 \frac{\mathrm{d}L_B}{\mathrm{d}t} + (\varepsilon_2 - \varepsilon_1)\frac{\mathrm{d}L_I}{\mathrm{d}t} = U_2 - U_1. \tag{5.3}$$

Overall mass balance for particles

The total volume V_B of particles in the bed is the sum for the two zones:

$$V_B = (1 - \varepsilon_2)L_I + (1 - \varepsilon_1)(L_B - L_I) = (1 - \varepsilon_1)L_B + (\varepsilon_1 - \varepsilon_2)L_I. \tag{5.4}$$

As V_B remains constant we have that $\mathrm{d}V_B/\mathrm{d}t = 0$, so that eqn (5.4), on differentiation, delivers a further relation linking the two interface velocities:

$$(1 - \varepsilon_1)\frac{\mathrm{d}L_B}{\mathrm{d}t} - (\varepsilon_2 - \varepsilon_1)\frac{\mathrm{d}L_I}{\mathrm{d}t} = 0. \tag{5.5}$$

The bed surface velocity

Summing eqns (5.3) and (5.5) yields the velocity of the bed surface u_{bs} during the transient response period:

$$u_{bs} = \frac{\mathrm{d}L_B}{\mathrm{d}t} = U_2 - U_1. \tag{5.6}$$

Thus, following a sudden change in fluid flux, the particles in the upper zone of the bed are predicted to travel at the constant velocity that is equal to this change. This is a notably simple relation, readily amenable to experimental verification. The total duration of the transient period T_T follows from relation (5.6):

$$T_T = \frac{L_2 - L_1}{U_2 - U_1}. \tag{5.7}$$

Experimental measurement of bed surface velocity

The relation of eqn (5.6) can be tested very easily using a video camera and recorder to monitor the bed surface position following a sudden change in fluid flux.

For reductions in the fluid flux to homogeneous liquid-fluidized beds, the rate of bed contraction has been found to follow exactly the predictions of eqn (5.6) throughout the entire transient period; the total time for completion of the changeover is given by eqn (5.7).

For increases in fluid flux the behaviour is less straightforward. The bed starts to respond in accordance with relation (5.6), the surface remaining quite flat and stable as for the case of a contracting bed. Then, some way into the expansion process, the bed surface starts to display eruptions, and its upward velocity falls below the predicted value, leading to a transient response time that can be significantly longer than that predicted by relation (5.7). This non-ideal behaviour becomes progressively more pronounced with increasing size of the initial step change in fluid flux (Figure 5.5).

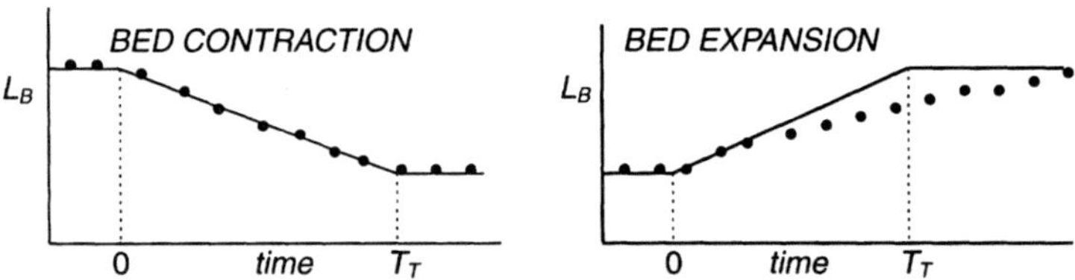

Figure 5.5 Comparison of idealized contraction/expansion predictions with experimental behaviour: model predictions, continuous lines; typical experimental data, points.

Gravitational instabilities

The departure from ideality of expanding beds can be fully explained in terms of gravitational instabilities (Figure 5.6). These occur as a consequence of the upper zone of the expanding bed being at a higher mean density than the lower zone. It is rather like having a vessel containing oil covered with a deep layer of water. With care such an arrangement is possible, but any disturbance is likely to result in globules of the oil detaching from the oil–water interface and rising through the higher density water layer.

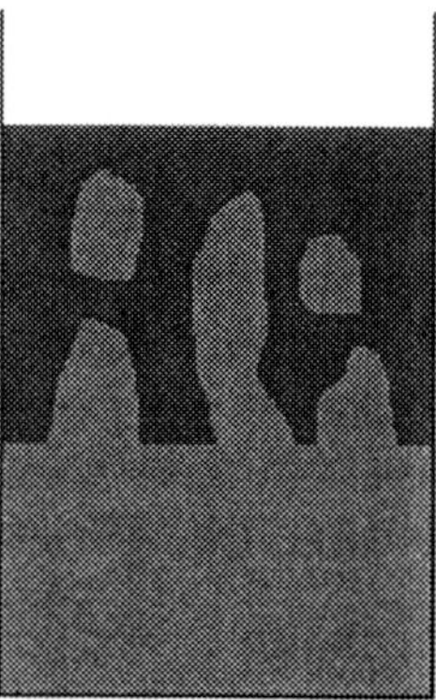

Figure 5.6 Gravitational instabilities in expanding, homogeneously fluidized beds.

This is precisely what happens in the expanding fluidized bed; the eruptions seen at the bed surface represent pockets of the relatively low-density suspension rising from the bottom zone. Particles above these rising pockets continue their unimpeded ascent at the constant velocity given by relation (5.6). It is only when the pockets reach the bed surface, and the homogeneity of the entire upper zone has been compromised, that the behaviour is seen to depart from the simple predictions.

Gravitational instabilities were not observed by Didwania and Homsy (1981) in *two-dimensional* water-fluidized beds. They found that eqn (5.6) held throughout for both contraction and expansion conditions, possibly as a result of the stabilizing influence of the bed geometry, which consisted of two sheets of glass placed close together to form a narrow rectangular slice in which the particle behaviour could be easily observed.

The kinematic-wave speed

Although the bed surface response represents the most obvious and easily measurable manifestation of the transient behaviour of homogeneously fluidized beds, it is the response of the other interface, that separating the two equilibrium zones, which provides a key component for a comprehensive analysis of the fluidized state. The velocity of this interface, $\mathrm{d}L_I/\mathrm{d}t$, follows immediately from the above relations: eqn (5.5) links it to the bed surface velocity, eqn (5.6), thereby yielding:

$$\frac{\mathrm{d}L_I}{\mathrm{d}t} = (1 - \varepsilon_1) \cdot \frac{U_2 - U_1}{\varepsilon_2 - \varepsilon_1}. \tag{5.8}$$

This relation describes the propagation of a finite discontinuity, or *shock*, separating two equilibrium states. The idealization implicit in its formulation is of particles in the bed making an instantaneous switch from one equilibrium condition to another as the shockwave passes over them. Only conservation of mass is involved in the analysis leading to eqn (5.8); inertial effects, which control the necessary deceleration of the moving particles to zero velocity, are not taken into account. The above analysis is thus in terms of a *kinematic* description of the fluidized state, and eqn (5.8) represents the velocity of a *kinematic shock* u_{KS}: $u_{KS} = \mathrm{d}L_I/\mathrm{d}t$.

The *kinematic-wave* velocity u_K is the limiting value of $\mathrm{d}L_I/\mathrm{d}t$ as the amplitude of the imposed fluid flux change $\Delta U(= U_2 - U_1)$ approaches the infinitesimal limit. Under these conditions eqn (5.8) yields:

$$\frac{\mathrm{d}L_I}{\mathrm{d}t} = u_{KS} \underset{\Delta U \to 0}{=} u_K = (1-\varepsilon)\cdot\frac{\mathrm{d}U}{\mathrm{d}\varepsilon}. \tag{5.9}$$

The derivative in eqn (5.9) relates to the steady-state, equilibrium expansion characteristics for homogeneous fluidization, which may be described by the empirical Richardson–Zaki relation, eqn (4.4): $U = u_t\varepsilon^n$. On evaluating the derivative, $\mathrm{d}U/\mathrm{d}\varepsilon$, from this expression, eqn (5.9) becomes:

$$u_K = nu_t(1-\varepsilon)\varepsilon^{n-1}. \quad \textit{Kinematic-wave velocity} \tag{5.10}$$

We shall see that this relation, which was first derived somewhat differently by Slis *et al.* (1959), plays a central role in the analysis of the fluidized state. It stipulates the velocity at which long-wavelength void fraction perturbation waves travel, always in the upward direction, through homogeneously fluidized beds. Perturbations represent an ever present reality in physical systems, created as a result of the imperfect nature of the fluid distributor and other imponderables.

Limitations to homogeneous behaviour

The behaviour described in this chapter has been the subject of extensive verification for homogeneous fluidization, in particular for liquid systems. The accuracy of its predictions for the shock and wave velocities is not in doubt: when the model works, it works exceptionally well. This raises the question concerning the many situations in which it fails completely.

Most gas-fluidized beds do not fluidize homogeneously. Nothing even approaching the ordered equilibrium behaviour and the ordered transition between equilibrium states described above is seen to occur in these cases. Instead, the behaviour is dominated by the marked inhomogeneities described in Chapter 1: bubbles of fluid coursing through the bed, carrying solids in their wake, and giving rise to intense chaotic mixing and large fluctuations in pressure. There is clearly something missing in the above analysis of ordered behaviour that fails to account for these gross disparities.

The clue to this problem has already been alluded to: it lies in the neglect of dynamic, inertial effects in the evaluation of the response of particles to the changes they experience in the interaction force. The fact that a kinematic description applies in some cases and not in others would therefore appear to relate to differences in the relative magnitude of dynamic to kinematic factors in different systems. These considerations are put on a quantitative footing in the following chapter.

References

Didwania, A.K. and Homsy, G.M. (1981). Rayleigh–Taylor instabilities in fluidized beds. *Ind. Eng. Chem. Fund.*, **20**, 318.

Gibilaro, L.G., Waldram, S.P. and Foscolo, P.U. (1984). A simple mechanistic description of the unsteady-state expansion of liquid-fluidized beds. *Chem. Eng. Sci.*, **39**, 607.

Slis, P.L., Willemse, Th. W. and Kramers, H. (1959). The response of the level of a fluidized bed to a sudden change in the fluidizing velocity. *Appl. Sci. Res.*, **A8**, 209.

6

A criterion for the stability of the homogeneously fluidized state

The dynamic-wave velocity

In the previous chapter, an idealized description of unsteady-state behaviour was seen to lead to remarkably simple, quantitative predictions for the response of a fluidized bed to changes in fluid flux; in particular to the velocity at which void fraction perturbations, generated at the distributor, travel up through the bed: the *kinematic-shock* velocity. This behaviour is observed experimentally for homogeneous fluidization typical of many liquid-fluidized systems. Implicit in the idealization is the notion of an instantaneous change in the velocity of the particles, from the equilibrium value they posses in the upper zone to zero, as the kinematic shock passes over them. The inertial response time for the particles must clearly be negligibly small for this condition to be applicable in practice. This raises the possibility of associating heterogeneous, bubbling fluidization with the failure of such systems to approximate sufficiently to this condition.

This idea was first proposed by Wallis (1962) in a research institute report. It appeared in the open literature, cast in somewhat more general terms, some years later with the publication of his book on two-phase flow (Wallis, 1969). A specific formulation of Wallis's criterion for homogeneous fluidization was soon to follow (Verloop and Heertjes, 1970). The inspiration appears to have been an analysis of road-traffic flow in terms of kinematic waves by Lighthill and Whitham (1955). In this work there is the description of an event, only too familiar to motorway drivers, which occurs when an accident, or some other partial interruption of the traffic flow, occurs some distance ahead. This gives rise to a 'traffic-concentration wave', which propagates back at a constant speed from the point of the obstruction. Drivers, who are for the most part unaware of the cause of the phenomenon, find that they are suddenly forced to slow down and move correspondingly closer together as the wave passes over them. The kinematic analysis disregards the inertial effects of breaking and acceleration, assuming these to be sufficiently rapid to have negligible influence on the overall behaviour. This assumption can sometimes prove over-optimistic: if the braking rate, for example, is insufficient to effect the required slow-down, perhaps as a result of the cars being too close together in the first place, then pile-ups will occur, signalling the breakdown of, among other things, the kinematic description.

The correspondence of this system to that of kinematic-wave propagation through fluidized beds, described in the previous chapter, is quite apparent: in that case it is the particles which have to slow down sufficiently rapidly as the kinematic-wave passes over them. All that is now required is for the limiting condition for a stable response to be identified, so that it becomes possible to see to what extent this differentiates between known instances of homogeneous and bubbling fluidization. This calls for some means of quantifying the inertial response time.

The speed at which *inertial* effects propagate through a system can be characterized in terms of another wave velocity, that of the *dynamic wave* – of which a common example is a pressure wave, which travels through air at the 'sonic' (i.e. dynamic-wave) velocity. A general theory then provides a remarkably simple means of quantifying, at least notionally, the condition for stable behaviour: *the velocity of the dynamic wave must be greater than that of the kinematic wave*. The basis for this condition will be illustrated in Chapter 8 with reference to the specific problem in hand. For now, we will simply attempt to apply it. For this we require an appropriate expression for the dynamic-wave velocity u_D in a fluidized bed.

The dynamic-wave velocity for the particle phase of a fluidized bed: the compressible fluid analogy

The particle phase of a homogeneously fluidized bed bears some resemblance to a compressible fluid. It can be 'compressed' by bearing down on the bed surface with a sieve (through which fluid, but not particles, may pass), removal of which results in the bed expanding back to its original height. For a small, localized compression of this type, which simply causes a layer of fluidized particles to be brought a little closer to an adjacent layer, the resemblance becomes more complete. We will draw on this perceived similarity in order to develop an expression for the dynamic-wave velocity through the particle phase (Foscolo and Gibilaro, 1984).

Consider first the case of a gas in an open cylinder fitted with a piston (Figure 6.1). A small, sudden upward displacement of the piston gives rise to compression of the gas immediately above it, and hence to a pressure wave which travels up the cylinder at the sonic (dynamic-wave) velocity u_D. Under adiabatic conditions, this velocity is given by:

$$u_D = \sqrt{\frac{\partial p}{\partial \rho_f}}. \qquad (6.1)$$

We now consider the counterpart experiment performed on the particle-phase of a homogeneously fluidized bed. The piston in this case is, in effect, the distributor (which acts like the sieve in the compression experiment

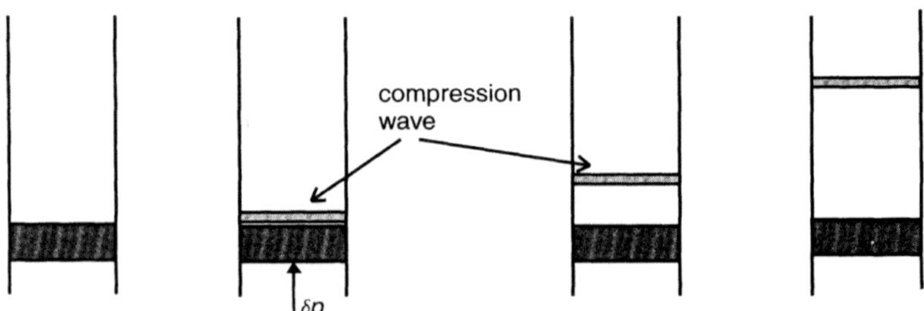

Figure 6.1 Dynamic wave creation in a gas.

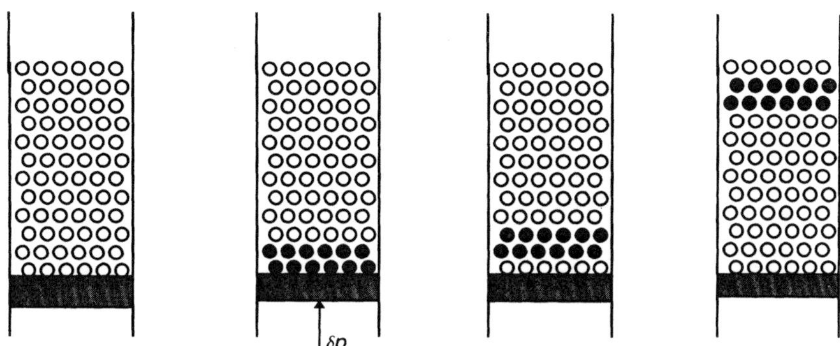

Figure 6.2 Dynamic wave creation in the particle phase of a fluidized bed.

referred to earlier). A small upward displacement of the piston/distributor 'compresses' the bottom particles, bringing the layer in contact with the distributor a little closer to the layer immediately above. This decrease in the local void fraction results in a net force on the second particle layer, causing it to accelerate upwards, restoring equilibrium below it, but imparting a net force on the layer immediately above, and so on. In this way a particle phase compression wave travels up the bed in a manner analogous to a sonic wave in a compressible fluid. The mechanism is quite different, however, to that for the compressible fluid, in that it is based on the dependence of void fraction on the net force acting on a fluidized particle.

This idealization of concentration-wave propagation through a particle phase is illustrated in Figure 6.2. It considers the particles to be arranged in regular, horizontal layers, so as to capture the essential feature of one-dimensional behaviour. We now see how this idealized arrangement enables an expression for the dynamic-wave velocity to be estimated by direct analogy with the expression of eqn (6.1) for the sonic velocity in a compressible fluid.

The pressure impulse δp applied to the frictionless piston/distributor translates into a net force on each of the particles that form the bottom layer of the bed:

$$\delta p \equiv \delta\{N_L f\}, \tag{6.2}$$

where N_L is the number of particles in a layer of unit area, and f is the net force experienced by each of these particles as a result of the local void fraction change brought about by the displacement of the piston.

The mean density ρ_{pp} of the particle phase is analogous to a gas density; the mass is provided by the particles, which are assumed to occupy the entire bed volume. ρ_{pp} therefore depends on the void fraction: $\rho_{pp} = (1-\varepsilon)\rho_p$. It is this density that changes with 'compression', the particles themselves being, in general, incompressible:

$$\delta\rho_{pp} \equiv \delta\{(1-\varepsilon)\rho_p\}. \tag{6.3}$$

By analogy with eqn (6.1), the dynamic-wave velocity for the particle phase thus becomes:

$$u_{\mathrm{D}} = \sqrt{\frac{\delta p}{\delta\rho_{pp}}} = \sqrt{\frac{\delta\{N_L f\}}{\delta\{(1-\varepsilon)\rho_p\}}}. \tag{6.4}$$

As the concentration wave may be regarded to be the consequence of solely an imposed perturbation in void fraction, eqn (6.4) may be written:

$$u_{\mathrm{D}} = \sqrt{\frac{\partial(N_L f)/\partial\varepsilon}{\partial(1-\varepsilon)\rho_p/\partial\varepsilon}} = \sqrt{-\frac{1}{\rho_p}\cdot\frac{\partial(N_L f)}{\partial\varepsilon}}. \tag{6.5}$$

It remains only to express N_L and f as functions of ε.

N_L can be estimated with the assumption that the void fraction on the horizontal planes that bisect the particle layers (Figure 6.2) is representative of the average void fraction in the bed. Thus:

$$N_L = 4(1-\varepsilon)/\pi d_p^2. \tag{6.6}$$

The net primary force f acting on a fluidized particle in equilibrium is simply the sum of drag, eqn (4.25), and effective weight, eqn (4.21):

$$f = \frac{\pi d_p^3}{6}\cdot(\rho_p - \rho_f)g\cdot\left[\left(\frac{U}{u_t}\right)^{\frac{4.8}{n}}\varepsilon^{-3.8} - \varepsilon\right]. \tag{6.7}$$

Note that on setting f to zero in eqn (6.7), we obtain the equilibrium relation: $U = u_t\varepsilon^n$.

The required derivative for eqn (6.5) can now be obtained from eqns (6.6) and (6.7):

$$\frac{\partial(N_L f)}{\partial\varepsilon} = N_L\frac{\pi d_p^3}{6}\cdot(\rho_p - \rho_f)g\cdot\left[-3.8\left(\frac{U}{u_t}\right)^{\frac{4.8}{n}}\varepsilon^{-4.8} - 1\right] + f\frac{\partial N_L}{\partial\varepsilon}. \tag{6.8}$$

On evaluating eqn (6.8) for a void fraction perturbation about the equilibrium condition ($f = 0$, $U = u_t\varepsilon^n$), with N_L given by eqn (6.6), and substituting this expression in eqn (6.5), we obtain the final relation for the dynamic-wave velocity:

$$u_{\mathrm{D}} = \sqrt{3.2gd_p(1-\varepsilon)(\rho_p - \rho_f)/\rho_p}. \tag{6.9}$$

This is a notably simple result, immediately available for any fluidized system. It has the same form as that proposed by Wallis (1962) on the basis of fluid-dynamic scaling considerations. The fact that the primary equilibrium force expression employed in the derivation, eqn (6.7), applies regardless of the flow regime provides for more generality than earlier formulations of stability criteria that assumed creeping flow conditions (Molerus, 1967; Verloop and Heertjes, 1970).

The stability criterion

Wallis's criterion for the stability of the state of homogeneous fluidization, $u_{\mathrm{D}} > u_{\mathrm{K}}$, can now be stated explicitly through eqns (5.10) and (6.9) for the kinematic- and dynamic-wave velocities respectively; it may be expressed in dimensionless form, $(u_{\mathrm{D}} - u_{\mathrm{K}})/u_{\mathrm{K}} > 0$:

$$\frac{1.79}{n}\cdot\left(\frac{gd_p}{u_t^2}\right)^{0.5}\left(\frac{\rho_p-\rho_f}{\rho_p}\right)^{0.5}\left(\frac{\varepsilon^{1-n}}{(1-\varepsilon)^{0.5}}\right) - 1 = \begin{array}{l} +\text{ve: homogeneous} \\ 0\text{: stability limit} \\ -\text{ve: bubbling} \end{array} \tag{6.10}$$

This expression is fully predictive. It enables the stability of any fluidized bed to be determined solely on the basis of the fundamental fluid and particle properties: ρ_f, μ_f, ρ_p, d_p (these properties deliver the required parameters n and u_t from eqns (4.5) and (2.17) respectively).

This stability criterion features prominently in later chapters, where it will be shown to provide reliable predictions of the stability of the homogeneously fluidized state for a vast range of experimentally tested systems. For now we simply illustrate its ability to differentiate between typical gas- and liquid-fluidized beds by means of a simple example: sand particles of diameter 200 μm and density 2500 kg/m^3 fluidized first by water (density 1000 kg/m^3, viscosity 10^{-3} Ns/m^2) and then by air (density 1.3 kg/m^3, viscosity 1.7×10^{-5} Ns/m^2).

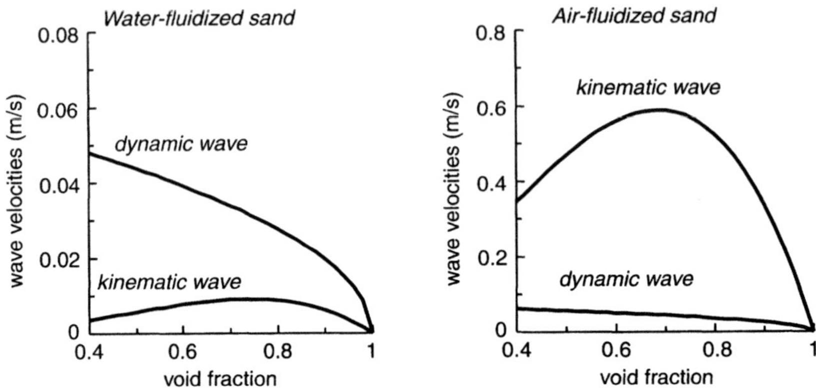

Figure 6.3 Stability of water- and gas-fluidized sand beds.

Figure 6.3 shows dynamic- and kinematic-wave velocities, eqns (6.9) and (5.10) respectively, for these two systems as functions of void fraction. It will be seen that for water fluidization, the dynamic-wave velocity is always well in excess of the kinematic-wave velocity, the reverse being the case for air fluidization – conforming to the well known behaviour of these systems, in which the water fluidization is always homogeneous and the air fluidization is always bubbling.

References

Foscolo, P.U. and Gibilaro, L.G. (1984). A fully predictive criterion for the transition between particulate and aggregate fluidization. *Chem. Eng. Sci.*, **39**, 1667.

Lighthill, M.J. and Whitham, G.B. (1955). On kinematic waves: II. A theory of traffic flow on long crowded roads. *Proc. R. Soc. (London)* **229A**, 317.

Molerus, O. (1967). Hydrodynamic stability of fluidized beds. *Chem. Eng. Technol.*, **39**, 341.

Verloop, J. and Heertjes, P.M. (1970). Shock waves as a criterion for the transition from homogeneous to heterogeneous fluidization. *Chem. Eng. Sci.*, **25**, 825.

Wallis, G.B. (1962). One-dimensional waves in two-component flow (with particular reference to the stability of fluidized beds). United Kingdom Atomic Energy Authority. Report AEEW-R162.

Wallis, G.B. (1969). *One-dimensional Two-Phase Flow*. McGraw-Hill.

7

The first equations of change for fluidization

A general formulation

The first published formulations of the governing equations for fluidization appeared in the scientific literature in the mid-1960s (Jackson, 1963; Murray, 1965; Pigford and Baron, 1965). This was a period of crucial importance for chemical engineering development, marking a change in emphasis away from applied, process-specific rules-of-thumb to the basic concepts embodied in the conservation laws for mass, momentum and energy transport. In that climate it was hardly surprising that a number of independent researchers should have been working towards the common goal of uncovering the fundamental laws governing the fluidization process. Some aspects of these initial investigations are described in *Research origins* in the opening pages of this book.

The analysis now to be presented represents a somewhat simplified generalization of formulations appearing around that time, which all arrived at the same conclusion regarding the stability of the state of homogeneous fluidization – a convergence that led to its almost universal acceptance. The main differences in the separate treatments concerned the manner in which the interaction force between the particle and fluid phases was formulated; we shall see that stability predictions are largely independent of such details.

The contemporaneous, unpublished work of Wallis (1962), referred to earlier, contained an additional term in the momentum equation that resulted in a rather different conclusion concerning the stability of the homogeneously fluidized state; this will be considered in some detail in the following chapters.

A one-dimensional, continuum description

In the analysis that follows, the particle phase of a fluidized bed is treated in some respects as though it were a continuum, or fluid. Thus the term *two-fluid model* is sometimes applied to this and related formulations. A differential control volume, Figure 7.1, is defined, from which equations specifying conservation of mass and momentum for the one-dimensional vertical flow of the fluid and particle phases may be written.

Both the particles and the fluid may be regarded as being incompressible. Although at first sight this appears inappropriate where the fluid is a gas, it represents a reasonable approximation, given that gas density changes resulting from the pressure drops encountered in normal fluidized bed applications remain relatively small.

The independent variables in the one-dimensional formulation are vertical height z and time t. The fluid occupies fraction ε of the control volume; fluid and particle velocities are u_f and u_p respectively; the fluid–particle interaction force per unit volume of suspension F_I is regarded as

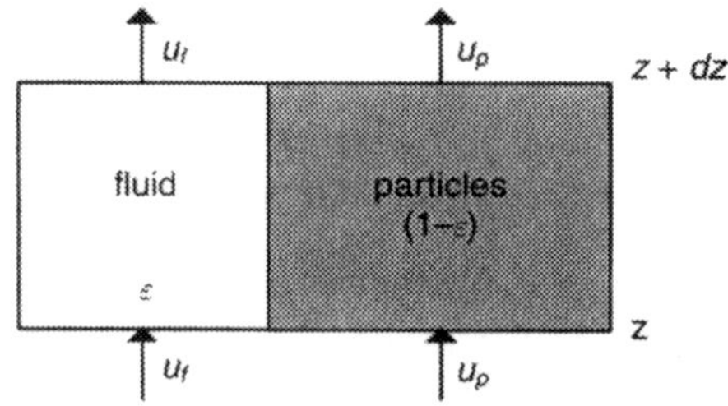

Figure 7.1 Control volume, of unit cross-sectional area, for a fluidized suspension.

being a function of ε, u_f and u_p. The particle phase of the control volume must relate to some average assembly of individual particles; in that sense it may be thought to contain a large number of them.

Conservation of mass

Mass balances on the incompressible fluid and particle components of the control volume, shown in Figure 7.1, yield:

$$\frac{\partial \varepsilon}{\partial t} + \frac{\partial}{\partial z}(\varepsilon u_f) = 0, \quad \textit{Fluid-phase} \tag{7.1}$$

$$-\frac{\partial \varepsilon}{\partial t} + \frac{\partial}{\partial z}\left[(1-\varepsilon)u_p\right] = 0. \quad \textit{Particle-phase} \tag{7.2}$$

Particle-phase mass balance
Rate of mass input – Rate of mass output = Rate of mass accumulation. (kg/m^{2}s)

Referring to Figure 7.1:

$$\left[u_p\rho_p(1-\varepsilon)\right]_z - \left[u_p\rho_p(1-\varepsilon)\right]_{z+dz} = \frac{\partial(dz(1-\varepsilon)\rho_p)}{\partial t},$$

$$\left[u_p(1-\varepsilon)\right]_z - \left[u_p(1-\varepsilon)\right]_{z+dz} = -dz\frac{\partial \varepsilon}{\partial t},$$

$$\left[u_p(1-\varepsilon)\right]_z - \left[\left[u_p(1-\varepsilon)\right]_z + dz\frac{\partial(u_p(1-\varepsilon))}{\partial z} + 0(dz^2)\right] = -dz\frac{\partial \varepsilon}{\partial t}.$$

Giving eqn (7.2) as $dz \to 0$.

The overall mass balance

This is obtained by summing eqns (7.1) and (7.2):

$$\frac{\partial}{\partial z}\left(\varepsilon u_f + (1-\varepsilon)u_p\right) = 0. \tag{7.3}$$

Equation (7.3) shows that the total flux (fluid plus particles, the quantity in brackets) remains constant, equal to that of the fluid entering the bed U_0. This result is a simple consequence of the particles and fluid being considered incompressible:

$$U_0 = \varepsilon u_f + (1-\varepsilon)u_p. \tag{7.4}$$

This equation links the fluid and particle velocities at all points in the bed; it shows that they are not independent of each other, so that the relative fluid–particle velocity u_{fp} may be expressed solely in terms of the particle velocity (or the fluid velocity):

$$u_{fp} = u_f - u_p = \frac{U_0 - (1-\varepsilon)u_p}{\varepsilon} - u_p = \frac{U_0 - u_p}{\varepsilon}. \qquad (7.5)$$

The fluid–particle interaction force per unit volume of the bed F_I may therefore be expressed solely as a function of u_p and ε, a conclusion which considerably simplifies the analysis that follows:

$$F_I = F_I(u_p, \varepsilon). \qquad (7.6)$$

Conservation of momentum

The control volume diagram shows velocities and volumetric concentrations for the two components; from this the momentum fluxes and accumulation rates may be readily evaluated. The forces acting on the fluid comprise: fluid–particle interaction $(-F_I dz)$, gravity $(-\varepsilon\rho_f g dz)$, and the net effect of fluid pressure, $(p(z) - p(z+dz))$. Because the particles are suspended in the fluid, the fluid pressure forces may be regarded as acting over the entire bed cross-section. The forces acting on the particles are simply fluid–particle interaction, acting in the opposite direction to that on the fluid $(F_I dz)$, and gravity $(-(1-\varepsilon)\rho_p g dz)$. Nothing analogous to a net pressure force is included in the particle momentum balance; such a term could be thought to arise from particle–particle collisions, in the same way that fluid pressure is transmitted as a result of molecular collisions. In so far as they were considered at all, such effects were regarded as being insignificant in the early formulations of the particle-phase equations.

On this basis, the momentum equations for the two components become:

$$\varepsilon\rho_f\left[\frac{\partial u_f}{\partial t} + u_f\frac{\partial u_f}{\partial z}\right] = -F_I - \varepsilon\rho_f g - \frac{\partial p}{\partial z}. \quad \textit{Fluid-phase} \qquad (7.7)$$

$$(1-\varepsilon)\rho_p\left[\frac{\partial u_p}{\partial t} + u_p\frac{\partial u_p}{\partial z}\right] = F_I - (1-\varepsilon)\rho_p g. \quad \textit{Particle-phase} \qquad (7.8)$$

Particle-phase momentum balance

Rate of momentum input − rate of momentum output + applied force = rate of momentum accumulation (momentum/s.m^2 = N/m^2)

Referring to the control volume diagram:

$$\left[u_p\rho_p(1-\varepsilon)u_p\right]_z - \left[u_p\rho_p(1-\varepsilon)u_p\right]_{z+dz}$$
$$+ F_I dz - (1-\varepsilon)\rho_p g dz = \frac{\partial}{\partial t}\left[(1-\varepsilon)dz\rho_p u_p\right]$$

giving:

$$\rho_p \frac{\partial}{\partial t}\left[u_p(1-\varepsilon)\right] + \rho_p \frac{\partial}{\partial z}\left[u_p^2(1-\varepsilon)\right] = F_I - (1-\varepsilon)\rho_p g.$$

On expanding the derivatives of the products in the terms on the left-hand side of the above equation, $u_p * (1-\varepsilon)$ and $u_p * u_p(1-\varepsilon)$ respectively, and applying continuity, eqn (7.2), we obtain eqn (7.8).

The particle-phase equations

Given that F_I may be regarded as a function solely of ε and u_p, eqn (7.6), it follows that the two particle-phase equations (7.2) and (7.8) represent a self-sufficient formulation, independent of the fluid velocity variable.

$$\frac{\partial \varepsilon}{\partial t} + u_p \frac{\partial \varepsilon}{\partial z} - (1-\varepsilon)\frac{\partial u_p}{\partial z} = 0, \quad \textit{Continuity} \tag{7.9}$$

$$(1-\varepsilon)\rho_p\left[\frac{\partial u_p}{\partial t} + u_p \frac{\partial u_p}{\partial z}\right] = F; \quad \textit{Momentum} \tag{7.10}$$

where F represents the net force (fluid–particle interaction plus gravity) acting on the particle phase:

$$F = F_I - (1-\varepsilon)\rho_p g. \tag{7.11}$$

Stability analysis

A trivial solution to eqns (7.9) and (7.10) is simply the steady-state condition, $\varepsilon = \varepsilon_0$ (a constant) and $u_p = 0$, which reduces the momentum equation, eqn (7.10), to $F = 0$. Given a constitutive expression for F, this relation delivers the constant, steady-state solution ε_0 for void fraction throughout the bed, a function solely of the fluid flux U_0. Such a solution represents the condition of homogeneous fluidization, and always satisfies the above particle-phase equations. The question now to be posed concerns the stability of this steady-state condition: is it sustainable in the face of small fluctuations in void fraction or particle velocity? Such fluctuations will always be present in an actual fluidized bed, if for no other reason than the impossibility of maintaining a perfect distribution of fluid at the entry region.

The linearized particle-phase equations

The full solution of eqns (7.9) and (7.10) to a small perturbation (in, say, the void fraction) imposed on a bed initially in the steady-state condition, $u_p = 0$ and $\varepsilon = \varepsilon_0$, would confirm the stability or otherwise of this condition: if the extent of the perturbation is found to increase with time, then the system must be deemed unstable, and *vice versa.* A much easier procedure, however, is to work with the *linearized* forms of the equations, which are always valid for *small* variations in the variables about the initial condition. The same test for stability may be applied much more easily in this case, as will soon become apparent. It should be pointed out, however, that the linear analysis reveals nothing about the final nature of an instability it identifies; all that it shows in this case is that perturbations *start* to grow, but whether this process leads to a fully bubbling bed or to some other, less-pronounced inhomogeneity, only a full non-linear analysis can reveal. This consideration features prominently in later chapters; for now attention will be focused solely on the question of linear stability.

By casting eqns (7.9) and (7.10) in terms of the deviation of void fraction ε^* from its steady-state level ε_0, $\varepsilon = \varepsilon_0 + \varepsilon^*$ (the other variable u_p is already a deviation about the steady-state value of 0), and eliminating terms that contain the product of two or more quantities which approach zero as u_p and ε^* approach zero, we obtain the linearized equations of change for the particle-phase:

$$(1-\varepsilon_0)\frac{\partial u_p}{\partial z} = \frac{\partial \varepsilon^*}{\partial t}, \quad \textit{Continuity} \tag{7.12}$$

$$(1-\varepsilon_0)\frac{\partial u_p}{\partial t} = \frac{u_p}{\rho_p}\mathrm{f}_{u_p} + \frac{\varepsilon^*}{\rho_p}\mathrm{f}_{\varepsilon}; \quad \textit{Momentum} \tag{7.13}$$

where f_{u_p} and f_{ε} are the partial derivatives of F with u_p and ε respectively, evaluated under equilibrium, steady-state conditions: $F = 0$:

$$\mathrm{f}_{u_p} = \left.\frac{\partial F}{\partial u_p}\right|_{F=0}, \quad \mathrm{f}_{\varepsilon} = \left.\frac{\partial F}{\partial \varepsilon}\right|_{F=0}. \tag{7.14}$$

Linearization of the particle-phase equations

Continuity. Eqn (7.9) in terms of deviation variables becomes:

$$\frac{\partial \varepsilon^*}{\partial t} + u_p \frac{\partial \varepsilon^*}{\partial z} - (1 - (\varepsilon_0 + \varepsilon^*)) \frac{\partial u_p}{\partial z} \doteq 0.$$

For small departures from equilibrium, the second term in this equation represents the product of two small quantities and can therefore be discounted; in the third term $\varepsilon_0 + \varepsilon^* \to \varepsilon_0$. This yields eqn (7.12).

Momentum. The right-hand side of eqn (7.10) is the net force acting on the particle phase, a non-linear function of the two variables u_p and ε. It may be approximated, for small deviations from the steady state, by the linear combination of the deviation variables obtained by truncating the Taylor expansion for F:

$$F(0 + u_p, \varepsilon_0 + \varepsilon^*) \approx F(0, \varepsilon_0) + u_p \left.\frac{\partial F}{\partial u_p}\right|_{F=0} + \varepsilon^* \left.\frac{\partial F}{\partial \varepsilon}\right|_{F=0}; \quad F(0, \varepsilon_0) = 0.$$

Inserting these relations in eqn (7.10), and linearizing the left-hand side in the same way as for the continuity equation, yields eqn (7.13).

The partial derivatives of the net force F

The partial derivative terms, eqn (7.14), represent constants, which may be readily evaluated given a constitutive relation for the fluid–particle interaction force. However, for the analysis that now follows it is sufficient to assume that the sign of both these constants is negative: simple qualitative considerations demonstrate that this must always be the case.

- Consider a stationary particle, $u_p = 0$, in equilibrium with a fluid flowing with velocity u_f; a small *increase* in particle velocity gives rise to a *reduction* in the fluid–particle relative velocity, $u_f - u_p$, and hence to a *reduction* in the net force on the particle: $\partial F/\partial u_p$ (i.e. f_{u_p}) *is always negative.*
- Consider the same particle, initially in equilibrium, subjected this time to a small *increase* in void fraction; this gives rise to a *reduction* in fluid velocity and hence a *reduction* in the fluid–particle relative velocity – leading to a *reduction* in the net force on the particle: $\partial F/\partial \varepsilon$ (i.e. f_ε) *is always negative.*

These conclusions are important in that they enable the following stability analysis to be performed without reference to a specific form for the fluid–particle interaction term in the momentum equation.

The linearized equations of change can be reduced to a single equation in void fraction by differentiating eqn (7.12) with respect to t, and eqn (7.13) with respect to u_p, thereby rendering the left-hand sides of these two equations identical; on then equating the right-hand sides and applying continuity, eqn (7.12), to eliminate the remaining term containing u_p, we obtain an equation solely in ε^*:

$$\frac{\partial^2 \varepsilon^*}{\partial t^2} + B\frac{\partial \varepsilon^*}{\partial t} + C\frac{\partial \varepsilon^*}{\partial z} = 0, \tag{7.15}$$

where:

$$B = -\frac{\mathrm{f}_{u_p}}{\rho_p(1-\varepsilon_0)}, \qquad C = -\frac{\mathrm{f}_\varepsilon}{\rho_p}. \tag{7.16}$$

As f_{u_p} and f_ε are negative quantities, it follows that B and C must always be positive.

The travelling wave solution

A solution to eqn (7.15) is provided by the void fraction perturbation wave having the form:

$$\varepsilon^* = \varepsilon_A \exp(at + ik(z - vt)), \tag{7.17}$$

where ε_A is the initial wave amplitude, a is the amplitude growth rate, k is the wave number ($k = 2\pi/\lambda$, where λ is the wavelength), and v is the wave velocity.

Eqn (7.17) describes the passage through the bed of a void fraction perturbation wave having an amplitude that either grows or decays with time according to the sign of the parameter a. These two possibilities signify instability and stability respectively. The wave solution thus provides a convenient route for establishing the conditions under which the homogeneously fluidized state is stable. It is also particularly straight forward to apply for this purpose, as we now see.

Writing the wave equation as the product of the time- and distance-dependent terms immediately delivers the partial derivative terms of eqn (7.15):

$$\varepsilon^* = \varepsilon_A \cdot \exp((a - ikv)t) \cdot \exp(ikz), \tag{7.18}$$

from which:

$$\frac{\partial \varepsilon^*}{\partial z} = ik\varepsilon^*, \quad \frac{\partial \varepsilon^*}{\partial t} = (a - ikv)\varepsilon^*, \quad \frac{\partial^2 \varepsilon^*}{\partial t^2} = (a - ikv)^2\varepsilon^*. \tag{7.19}$$

Inserting these expressions into eqn (7.15) yields the complex algebraic relation:

$$(a^2 - k^2v^2 + Ba) + (-2akv - Bkv + Ck)i = 0. \tag{7.20}$$

Equating the real and imaginary terms of eqn (7.20) to zero yields:

$$a = \frac{C - Bv}{2v}, \tag{7.21}$$

$$k^2 = \frac{C^2 - B^2v^2}{4v^4}. \tag{7.22}$$

Instability of the homogeneously fluidized state

Eqns (7.21) and (7.22) provide a clear answer to the stability question for the rather general formulation of the problem considered above. k, the wave number, is a real, positive quantity, so k^2 must be positive. Equation (7.22) thus yields the condition that $C > Bv$; and hence, from eqn (7.21), that a must always be positive. The simple conclusion arising from the analysis is that *the homogeneously fluidized state is intrinsically unstable.*

This conclusion provided an emphatic justification for the phenomenon of bubbles in gas-fluidized beds. As these beds represented the most widespread and important industrial applications of fluidization technology, it is not difficult to appreciate the considerable interest generated in the mid-1960s by the diverse analyses (by different authors, employing different fluid–particle interaction force expressions) that arrived at essentially the same conclusion. The inconvenient fact that liquid-fluidized beds appeared to manifest stable, homogeneous behaviour did little to detract from its almost universal acceptance.

At this point the reader may well be feeling perplexed at the striking inconsistency of this conclusion with those arrived at in the two preceding chapters. In Chapter 5, a simple analysis of *stable*, homogeneous fluidization led to predictions of bed behaviour in good agreement with experimental observations of liquid-fluidized systems. In Chapter 6, a criterion for distinguishing stable from unstable (bubbling) fluidization was derived and shown to differentiate typical, bubbling gas fluidization from typical, homogeneous liquid fluidization. Now it appears that these findings, and other successful predictions relating to the existence of a stable, homogeneous state, are incompatible with a seemingly general formulation of the mass and momentum conservation laws.

The justification advanced at the time for disregarding the apparently stable behaviour of liquid-fluidized systems relates to the limitations of the linear analysis referred to earlier: linear instability only guarantees that perturbations *start* to grow; where they eventually end up is anybody's guess. Various plausible hypotheses were proposed to explain why apparently homogeneous liquid beds are in fact unstable. One drew on experimental observations in certain liquid-fluidized systems of high voidage bands propagating upwards through the bed under certain conditions – which could represent the final outcome of perturbation growth stopping well short of bubble formation (Jackson, 1985). This phenomenon is discussed in some detail in Chapters 9–12. Another hypothesis allowed for completely void bubble formation, but postulated a greatly reduced bubble size in liquid systems, of the order of particle size, and hence effectively undetectable (Harrison *et al.*, 1961). Such arguments are difficult to counter, and for a number of years the notion of the intrinsic instability of the homogeneously fluidized state held sway. This state of affairs may well have continued indefinitely had it not been for the discovery of the remarkable behaviour of gas fluidized *fine* powders, which served to convince all but the most devoted adherents to the intrinsic instability concept that something was missing in the accepted theory.

The next chapter starts with an account of the essential features of fine-powder gas fluidization. It then goes on to justify the inclusion of an additional fluid-dynamic term in the equations of change for the fluidized particle phase, which leads to the satisfactory resolution of the stability problem.

References

Harrison, D., Davidson, J.F. and de Kock, J.W. (1961). On the nature of aggregate and particulate fluidization. *Trans. Inst. Chem. Eng.*, **39**, 202.

Jackson, R. (1963). The mechanics of fluidized beds: Part 1: The stability of the state of uniform fluidization. *Trans. Inst. Chem. Eng.*, **41**, 13.

Jackson, R. (1985). Hydrodynamic stability of fluid–particle systems. In: *Fluidization*, 2nd edn (J.F. Davidson, R. Clift and D. Harrison, eds). Academic Press.

Murray, J.D. (1965). On the mathematics of fluidization. 1. *J. Fluid Mech.*, **21**, 465.

Pigford, R.L. and Baron, T. (1965). Hydrodynamic stability of a fluidized bed. *Ind. Eng. Fund.*, **4**, 81.

Wallis, G.B. (1962). One-dimensional waves in two-component flow (with particular reference to the stability of fluidized beds). United Kingdom Atomic Energy Authority. Report AEEW-R162.

8

The particle bed model

Fine-powder gas fluidization

In the previous chapter, a general formulation of the equations of change was shown to lead to the conclusion that the state of homogeneous fluidization is intrinsically unstable. This appears to conflict with observations of apparently homogeneous fluidization in liquid-fluidized beds, but ambiguities in the behaviour of some of these systems have been cited to support the notion that, although the instability they manifest is far less extreme than is the case for gas fluidization, they are nevertheless also unstable in a formal sense.

The death knell of the 'intrinsic instability' hypothesis was sounded with the disclosure of the expansion characteristics of gas fluidized 'fine' powders (d_p in the range of approximately 40–100 μm). These systems were found to start expanding, with increasing gas flux, in an unambiguously homogeneous manner up to a critical, well-defined value of void fraction ε_{mb}, and thereafter in the bubbling mode (Geldart, 1973). The experimental observations leave no room for doubt in the matter; below the critical gas flux U_{mb} the bed appears absolutely stable, expanding to perhaps twice or more of its original height with increasing gas flux U_0, while maintaining a completely flat, undisturbed surface. At the critical transition point (U_{mb}, ε_{mb}) marked instabilities can sometimes be observed, with the bed surface exhibiting violent, large-amplitude oscillations; in Chapter 14 these will be shown to be a predictable consequence of system non-linearities, which the following formulation describes. At a slightly higher gas flux these oscillations (if they occur) vanish, and small gas bubbles can be observed breaking through the bed surface, as is the case with 'normal' gas fluidization just beyond the minimum fluidization point (U_{mf}, ε_{mf}). Further increases in gas flux give rise to the familiar phenomenon, for gas systems, of a freely bubbling bed. The major effect of the reduction in particle size is thus to separate the minimum fluidization point from the minimum bubbling point, interposing between these two critical conditions a region of unambiguously stable, homogeneous expansion. The following formulation of the equations of change for fluidization will be shown to provide a quantitative explanation for this phenomenon.

The primary force interactions

The development to be described in the following sections focuses on the role of individual fluidized particles, their force interactions with the surrounding fluid, and the effect on the net force experienced by a particle of *approaching* particle-concentration (or void fraction) perturbations – which, as we shall see, impart an effective *elasticity* to the particle phase: it is this latter factor that gives rise to an additional term in the momentum equation for the particle phase, rendering it capable of accommodating stable, homogeneous behaviour (Foscolo and Gibilaro, 1987). First, however, we set down the *primary forces*, which were evaluated in Chapter 4 for the special case of a fluidized particle under conditions of equilibrium.

Under *equilibrium* conditions the net primary force f_0 comprises drag, eqn (4.25), and the 'effective particle weight' (the net effect of gravity and buoyancy), eqn (4.21):

$$f_0 = \frac{\pi d_p^3}{6}(\rho_p - \rho_f)g\left(\frac{U_0}{u_t}\right)^{\frac{4.8}{n}}\varepsilon^{-3.8} - \frac{\pi d_p^3}{6}(\rho_p - \rho_f)g\varepsilon. \tag{8.1}$$

Setting $f_0 = 0$ in eqn (8.1) yields the steady-state expansion law: $U_0 = u_t\varepsilon^n$.

Under *non-equilibrium* conditions, two modifications must be made to the net primary force expression of eqn (8.1). First of all, the drag force must be expressed in terms of the relative fluid–particle velocity u_{fp}, given by eqn (7.5), rather than the steady-state fluid velocity, which is equal to U_0/ε. This involves replacing U_0 in eqn (8.1) with εu_{fp} ($= U_0 - u_p$, see eqn (7.5)). And secondly, the buoyancy contribution must retain its general, non-equilibrium form: eqn (4.17). On this basis, the net primary force f on a fluidized particle becomes:

$$f = \frac{\pi d_p^3}{6}\left[(\rho_p - \rho_f)g\left(\frac{U_0 - u_p}{u_t}\right)^{\frac{4.8}{n}}\varepsilon^{-3.8} - \frac{\partial p}{\partial z} - \rho_p g\right]. \tag{8.2}$$

The number N_V of particles per unit volume of bed is simply:

$$N_V = \frac{6(1-\varepsilon)}{\pi d_p^3}, \tag{8.3}$$

leading to the net primary force F per unit volume of bed:

$$\begin{aligned} F &= F_d + F_b + F_g \\ &= (1-\varepsilon)\left[(\rho_p - \rho_f)g\left(\frac{U_0 - u_p}{u_t}\right)^{\frac{4.8}{n}}\varepsilon^{-3.8} - \frac{\partial p}{\partial z} - \rho_p g\right]. \end{aligned} \tag{8.4}$$

We now turn to the key concept of the *particle bed* model, the introduction of which gives rise to a quite different conclusion for system stability to that reported in the previous chapter.

Fluid-dynamic elasticity of the particle phase

The above expressions for the net primary force on a particle and the particle phase, eqns (8.2) and (8.4) respectively, have been arrived at largely on the basis of equilibrium, steady-state considerations. Questions arise, however, concerning the unsteady state: to what extent do other significant force mechanisms come in to play in this case, and how may these be quantified?

In order to go some way towards addressing these questions, consider the following idealized description of a particle-concentration (or void fraction) perturbation imposed on a fluidized bed initially in the homogeneous, equilibrium state. Such perturbations rise vertically through fluidized beds, where they are always present, due, as we saw in the previous chapter, to such things as the imperfect nature of the fluid distributor at the base. If viewed with a scale of scrutiny of the order of particle size, these rising particle-concentration perturbations would be seen to alter the fluid flow-field for some small distance ahead of them; particles above will therefore sense, and start to respond to, the changing flow conditions before the particle-concentration disturbance itself actually reaches them, the effect increasing as the disturbance gets closer. What has just been described represents an effectively *elastic* response, the transmitted force *increasing* as the particle phase is '*compressed*' (or decreasing as the particle-phase is expanded). This mechanism is not included in the early treatments of fluidization dynamics described in the previous chapter, and represented by eqns (7.9) and (7.10).

In order to try to quantify the above elastic effect, we now consider the usual situation, encountered above and throughout the previous chapter, in which the net force on a fluidized particle (or element of the particle-phase) is regarded to be a function of, among other things, particle concentration α. This seems a more appropriate variable, when considering particle-phase elasticity, than void fraction ε, with which it is, of course, readily interchangeable:

$$\alpha = 1 - \varepsilon. \tag{8.5}$$

On this basis, the effect on the force experienced by a fluidized particle as a result of another particle approaching close to it from below (effectively the situation described in the previous paragraph) can be expressed somewhat differently, as will now be described.

Particle concentration, by definition, must relate to a volume of finite size, a *region of influence*, which is certainly larger than a single particle. This implies that a concentration-dependent force on a particular particle can be affected by, say, another particle entering its region of influence (Figure 8.1), thereby increasing the particle concentration in that region: direct collision between particles is therefore not a necessary requirement for transfer of momentum between them.

This mechanism is qualitatively similar to the more fundamental one described previously in terms of alterations to the fluid flow-field brought about by an approaching particle-concentration perturbation: both describe how fluid-dynamic forces may be transmitted between fluidized particles. The latter description, however, readily lends itself to simple quantification.

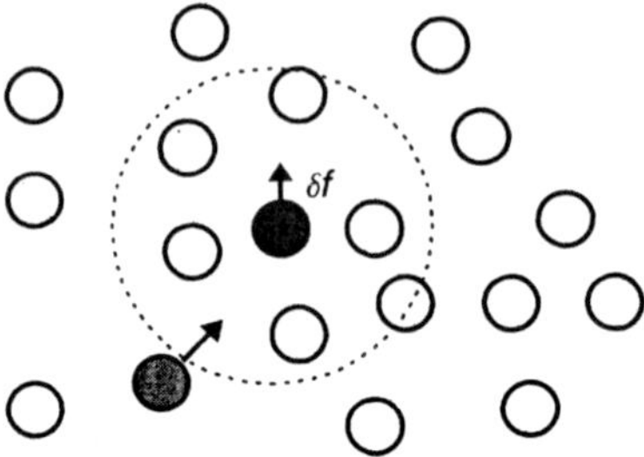

Figure 8.1 The 'region of influence' for a fluidized particle.

The penetration distance

Consider now the one-dimensional formulation, in which concentration perturbations extend across a horizontal plane and travel vertically upwards through the bed. This situation conforms to a considerable degree to that observed in actual, homogeneously-fluidized systems. We have seen that, under unsteady-state conditions, the *effective particle concentration*, on the basis of which the net force on a particle is to be evaluated, must relate to a finite region associated with the particle. For the one-dimensional case this region corresponds to a penetration distance Δz in the vertical direction, measured downwards from the particle centre. This finite region of influence must be preserved in the differential equation description of momentum transfer so that the ability of a particle to respond to approaching concentration perturbations, before they actually arrive at the particle centre line, is not lost. This can be achieved by means of the following definition for effective particle concentration α_e:

$$\alpha_e = \alpha - \Delta z \frac{\partial \alpha}{\partial z}. \tag{8.6}$$

Equation (8.6) reverts to the trivial form, $\alpha_e = \alpha$, under equilibrium conditions. In the presence of a concentration gradient, however, it provides an estimate of particle concentration in the finite region of influence associated with the particle in question.

On this basis, the dependence of net force on particle concentration may be written:

$$f(\alpha_e) = f\left(\alpha - \Delta z \frac{\partial \alpha}{\partial z}\right) \approx f(\alpha) - \Delta z \frac{\partial f}{\partial \alpha} \frac{\partial \alpha}{\partial z}. \tag{8.7}$$

Closure requires solely an estimate for the penetration distance Δz.

An estimate for penetration distance

We now show how a working estimate for penetration distance Δz can be obtained by assuming the idealized geometric arrangement introduced in Chapter 6 for the homogeneously suspended particles. Although no actual fluidized bed, or other dispersed system, could be expected to conform exactly to any fixed configuration, this one provides a route to what appears a reasonable first estimate, the derived result supporting intuitive physical considerations, which suggest the order of a particle diameter for Δz (see Figure 8.2).

The fluidized bed in equilibrium is thus considered to consist of particles arranged in regular horizontal layers.

Each layer contains N_L particles per unit area. A representative volume element for this bed is provided by the volume included between two horizontal planes of unit area that bisect adjacent particle layers. This volume contains N_L particles (in fact, $2N_L$ half-particles). If the distance between adjacent layers is θ, then the average particle concentration in the bed is given by:

$$\alpha = \pi d_p^3 N_L / 6\theta. \tag{8.8}$$

A horizontal plane through the centre of a layer only passes through particles in that layer, bisecting them. If we stipulate that the particle concentration evaluated on such a plane is representative of the average particle concentration in the bed as a whole, then we have:

$$\alpha = \pi d_p^2 N_L / 4. \tag{8.9}$$

This arrangement ensures a narrow distribution of particle concentration across horizontal planes drawn between adjacent layers, the maximum

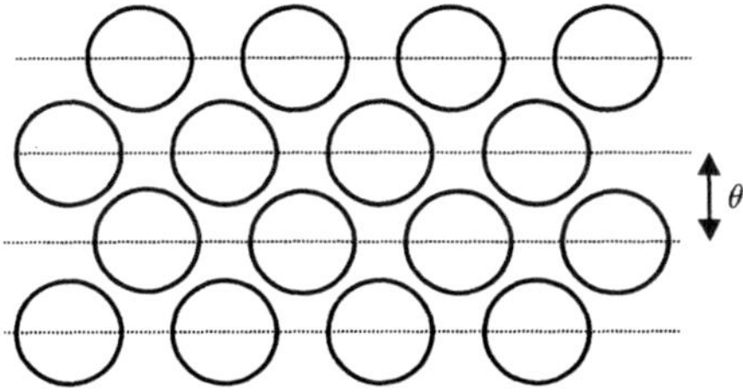

Figure 8.2 Idealized particle layer description of a homogeneously fluidized bed.

deviation approximating to 10 per cent of the mean α. Equating the two expressions for particle concentration, eqns (8.8) and (8.9), yields the distance θ between adjacent layers:

$$\theta = 2d_p/3. \tag{8.10}$$

This 'horizontal layer' description of a fluidized bed effectively furnishes an estimate for the penetration distance Δz, which defines the region over which changes in particle concentration affect the force on a particular particle. Certainly, the effect of a vertical displacement of a particle layer will be felt by the particles in the layer immediately above, and it would seem reasonable to assume that disturbances further removed would have effect only in so far as they successively displace the intermediate layers. This all suggests that an appropriate estimate for penetration distance is simply the layer spacing:

$$\Delta z = \theta = 2d_p/3. \tag{8.11}$$

On the basis that the overall behaviour of a bed is insensitive to the specific particle orientation, we will use this conclusion as a generally applicable working hypothesis. Different assemblies for the uniform suspension would lead to minor quantitative changes in the final result. Thus the general expression for the total force f^+ experienced by a fluidized particle may be written:

$$f^+ = f + f_{\Delta z} = f - \left(\frac{2d_p}{3}\frac{\partial f}{\partial \alpha}\right)\frac{\partial \alpha}{\partial z}, \tag{8.12}$$

where f, a function of α and the velocity of the fluid relative to the particle, is the net primary force evaluated purely on the basis of steady-state considerations; and the second term takes into account the elastic effect brought about by rising concentration perturbations approaching the particle from below.

The elastic modulus of the particle phase

A unit volume of bed contains $6\alpha/\pi d_p^3$ particles, so that the total net force F^+ acting on the particle phase in a unit volume of suspension becomes:

$$F^+ = \frac{6\alpha}{\pi d_p^3}\left(f - \left(\frac{2d_p}{3}\frac{\partial f}{\partial \alpha}\right)\frac{\partial \alpha}{\partial z}\right). \tag{8.13}$$

This defines the force term that is to appear in the particle-phase momentum equation. Closure requires no more than the expression for the net primary force f, provided here by eqn (8.2). Before effecting this closure, a general interpretation of the elastic component of the total force may prove useful.

Eqn (8.13) may be written:

$$F^+ = F - E\frac{\partial \alpha}{\partial z}, \tag{8.14}$$

where $F^+ = F$ under equilibrium conditions. The second term in eqn (8.14) is analogous to a pressure gradient (the *particle-pressure* gradient), $\partial p_p/\partial z$, and E may be taken to represent the elastic modulus of the particle phase:

$$\frac{\partial p_p}{\partial z} = E\frac{\partial \alpha}{\partial z}, \tag{8.15}$$

$$E = \frac{4\alpha}{\pi d_p^2}\cdot\frac{\partial f}{\partial \alpha} = -\frac{4(1-\varepsilon)}{\pi d_p^2}\cdot\frac{\partial f}{\partial \varepsilon}. \tag{8.16}$$

We are considering the case of a fluidized bed, initially in equilibrium, responding to a rising particle-concentration perturbation. Under these circumstances, the derivative in the elasticity expression, eqn (8.16), is to be evaluated from the equilibrium relation for the net force, eqn (8.1), at $f_0 = 0$ (where $U_0 = u_t\varepsilon^n$)

$$\begin{aligned}\frac{\partial f_0}{\partial \varepsilon} &= \frac{\pi d_p^3}{6}(\rho_p - \rho_f)\left[-3.8\left(\frac{U_0}{u_t}\right)^{\frac{4.8}{n}}\varepsilon^{-4.8} - 1\right] \\ &\underset{f_0=0}{=} -4.8\frac{\pi d_p^3}{6}(\rho_p - \rho_f)g;\end{aligned} \tag{8.17}$$

giving for the particle-phase elasticity:

$$E = 3.2gd_p(1-\varepsilon)(\rho_p - \rho_f). \tag{8.18}$$

The dynamic-wave velocity

The dynamic-wave velocity for the particle phase, u_D, may be expressed in terms of E:

$$u_D = \sqrt{E/\rho_p} = \sqrt{3.2gd_p(1-\varepsilon)(\rho_p - \rho_f)/\rho_p}. \tag{8.19}$$

This expression for u_D is precisely that obtained in Chapter 6, where it was derived by drawing on the analogy of the particle phase as a compressible fluid. The arguments presented above clarify the precise mechanism that gives rise to its inclusion in the momentum equation.

Note that for the case of gas fluidization, where $\rho_p \gg \rho_f$, eqn (8.19) reduces to:

$$u_D = \sqrt{3.2 g d_p (1 - \varepsilon)}. \tag{8.20}$$

The particle bed model

We are now in a position to write down the closed formulation of the conservation equations for the particle and fluid phases that defines the one-dimensional *particle bed* model. The particle phase momentum equation, derived in Chapter 7, adopts the net primary force term F of eqn (8.4) and is augmented by the elasticity term $F_{\Delta z}$, which, in terms of the dynamic-wave velocity, becomes: $\rho_p u_D^2 \partial\varepsilon/\partial z$.

The particle-phase equations

$$-\frac{\partial \varepsilon}{\partial t} + \frac{\partial}{\partial z}\left[(1-\varepsilon)u_p\right] = 0, \tag{8.21}$$

$$\begin{aligned}(1 - \varepsilon)\rho_p\left[\frac{\partial u_p}{\partial t} + u_p\frac{\partial u_p}{\partial z}\right]\\ = F_d - (1-\varepsilon)\frac{\partial p}{\partial z} - (1-\varepsilon)\rho_p g + \rho_p u_D^2 \frac{\partial \varepsilon}{\partial z}.\end{aligned} \tag{8.22}$$

The fluid-phase equations

$$\frac{\partial \varepsilon}{\partial t} + \frac{\partial}{\partial z}(\varepsilon u_f) = 0, \tag{8.23}$$

$$\varepsilon\rho_f\left[\frac{\partial u_f}{\partial t} + u_f\frac{\partial u_f}{\partial z}\right] = -F_d - \varepsilon\frac{\partial p}{\partial z} - \varepsilon\rho_f g. \tag{8.24}$$

Note that the second term on the right-hand side of the fluid momentum equation (8.24), $-\varepsilon(\partial p/\partial z)$, comprises the sum of two terms: the buoyant interaction with the particles in the control volume, $(1-\varepsilon)\partial p/\partial z$, and the net surface force across the control volume boundaries, $-\partial p/\partial z$.

Gas fluidization: the single-phase particle bed model

The stability of the homogeneously fluidized state is analysed in terms of the full set of system equations, eqns (8.21)–(8.24), in Chapter 11. For the case of *gas fluidization*, however, where $\rho_p \gg \rho_f$, terms in the fluid momentum equation that are proportional to fluid density will be negligible compared to the drag and pressure gradient terms, which are involved in supporting the fluidized particles. Equation (8.24) then reduces to:

$$F_d = -\varepsilon \frac{\partial p}{\partial z}, \tag{8.25}$$

a relation which enables the fluid pressure gradient term in eqn (8.22) to be replaced, thereby decoupling the fluid- and particle-phase equations:

The particle-phase momentum equation for $\rho_p \gg \rho_f$

$$\begin{aligned}(1-\varepsilon)\rho_p\left[\frac{\partial u_p}{\partial t} + u_p\frac{\partial u_p}{\partial z}\right] &= F + \rho_p u_{\mathrm{D}}^2 \frac{\partial \varepsilon}{\partial z},\\ F &= \frac{F_d}{\varepsilon} - (1-\varepsilon)\rho_p g.\end{aligned} \tag{8.26}$$

Equations (8.21) and (8.26) now represent a closed formulation that is valid for all cases of gas fluidization, and also serves as a working approximation for the liquid fluidization of relatively dense particles.

A different decoupling procedure was employed in the original *particle bed* model formulation (Foscolo and Gibilaro, 1987). There the fluid pressure gradient was approximated by its equilibrium value, eqn (4.18), leading to a somewhat different primary interaction term to that of eqn (8.26). Both procedures lead to the same stability criterion (derived below), but the present approximation has the advantage of being fully consistent with the complete formulation, eqns (8.21)–(8.25), leading, as we shall see, to identical solutions for the case of gas fluidization, $\rho_p \gg \rho_f$.

Stability analysis

As was the case for the formulation considered in the previous chapter, eqns (8.21) and (8.26) are satisfied by the trivial solution: $u_p = 0$, $\varepsilon = \varepsilon_0$, representing steady-state, homogeneous fluidization. The problem once again is to determine under what conditions, if any, such a solution is stable. We now proceed as in the previous chapter: eqns (8.21) and (8.26) are linearized, and the particle velocity variable terms eliminated in exactly the manner applied to eqns (7.9) and (7.10); this time, however, we have a constitutive expression for the net primary force F, leading to specific forms for the (always negative) partial derivative terms, f_{u_p} and f_{ε}, defined in eqn (7.14):

$$\mathrm{f}_{\varepsilon} = -4.8\rho_p g \frac{1-\varepsilon_0}{\varepsilon_0}, \qquad \mathrm{f}_{u_p} = \frac{\mathrm{f}_{\varepsilon}(1-\varepsilon_0)}{u_{\mathrm{K}}}; \tag{8.27}$$

the linearized net primary force thus becomes:

$$F = \mathrm{f}_{\varepsilon}\varepsilon^* + \mathrm{f}_{u_p} u_p = -D\rho_p\left(u_{\mathrm{K}}\varepsilon^* + (1-\varepsilon_0)u_p\right), \tag{8.28}$$

where D is a positive quantity, and u_{K} is the kinematic-wave velocity (previously derived independently in Chapter 5), which arises naturally from the formulation of fluid–particle interaction adopted in the model:

$$D = \frac{4.8g(1-\varepsilon_0)}{u_{\mathrm{K}}\varepsilon_0}, \tag{8.29}$$

$$u_{\mathrm{K}} = u_t n(1-\varepsilon_0)\varepsilon_0^{n-1}. \tag{8.30}$$

On eliminating u_p from the linearized particle-phase equations as before, we obtain:

$$\frac{\partial^2\varepsilon^*}{\partial t^2} - u_{\mathrm{D}}^2\frac{\partial^2\varepsilon^*}{\partial z^2} + D\left(\frac{\partial\varepsilon^*}{\partial t} + u_{\mathrm{K}}\frac{\partial\varepsilon^*}{\partial z}\right) = 0. \tag{8.31}$$

This equation, which contains the two key wave velocities, u_{D} and u_{K}, has an additional term (the second) to that of the original formulation, eqn (7.15). It is satisfied by the same travelling wave solution, eqn (7.17).

On evaluating the partial derivatives from the wave equation, eqn (7.17), and inserting them in eqn (8.31), we obtain the following expressions for the wave amplitude growth rate a and the wave number k:

$$a = \frac{D}{2v}(u_{\mathrm{K}} - v), \tag{8.32}$$

$$k^2 = \frac{D^2}{4v^2}\left(\frac{u_{\mathrm{K}}^2 - v^2}{v^2 - u_{\mathrm{D}}^2}\right). \tag{8.33}$$

Equation (8.33) reveals immediately the velocity bounds for a perturbation wave: short waves ($k \to \infty$) and long waves ($k \to 0$)

approach respectively the dynamic- and kinematic-wave speeds, u_D and u_K.

The stability criterion

Eqn (8.32) indicates that stable, homogeneous fluidization occurs when the velocity v of the perturbation wave is greater than that of the kinematic wave u_K; a is then negative so that perturbation amplitudes decay with time. For this condition, $v > u_K$, the numerator in eqn (8.33) is negative; this means that for real values of the wave number $k\,(k^2 > 1)$ the denominator in eqn (8.33) must also be negative. Thus we have: $u_D > v > u_K$, and hence the condition for stable, homogeneous fluidization may be written:

$$u_D > u_K. \quad \textit{Stable fluidization} \tag{8.34}$$

The wave number k is also real when both the numerator and denominator in eqn (8.33) are positive; in this case we have: $u_K > v > u_D$. Under these conditions eqn (8.32) reveals the perturbation growth rate a to be positive, indicating unstable, heterogeneous fluidization, and hence the condition:

$$u_K > u_D. \quad \textit{Unstable fluidization} \tag{8.35}$$

Implicit in the stability conditions of eqns (8.34) and (8.35) is the fact that the wave velocity v is always bounded by u_D and u_K.

The full stability criterion is therefore given by:

$$u_D - u_K \begin{array}{ll} +\text{ve} & \textit{Stable: homogeneous fluidization} \\ = 0 & \textit{Stability limit: } \varepsilon = \varepsilon_{mb} \\ -\text{ve} & \textit{Unstable: bubbling fluidization} \end{array} \tag{8.36}$$

Eqn (8.36) is the statement of the general Wallis (1962, 1969) criterion for fluidized bed stability. The specific forms for the dynamic- and kinematic-wave velocities arising from the model formulation, eqns (8.19) and (8.30), yield the closed form of this criterion, which was derived indirectly in Chapter 6, and expressed in eqn (6.10).

The application to any fluidized system could not be easier: it requires only a knowledge of the basic particle and fluid properties (ρ_p, d_p, ρ_f, μ_f) from which values of u_D and u_K, and hence the stability condition, follow explicitly. Table 8.1 summarizes the calculation procedure.

Table 8.1 Summary of fluidized bed stability determination

1 Input system parameter values	ρ_p, d_p, ρ_f, μ_f
2 Evaluate the Archimedes number Ar	$Ar = gd_p^3\rho_f(\rho_p - \rho_f)/\mu_f^2$
3 Evaluate u_t and n	$u_t = \left[-3.809 + (3.809^2 + 1.832Ar^{0.5})^{0.5}\right]^2 \cdot \mu_f/(\rho_f d_p)$ from Eqn (2.17) $n = \dfrac{4.8 + 0.1032Ar^{0.57}}{1 + 0.043Ar^{0.57}}$ from Eqn (4.5)
4 Set ε at minimum fluidization value	$\varepsilon = 0.4$
5 Evaluate u_D and u_K	$u_D = \sqrt{E/\rho_p} = \sqrt{3.2gd_p(1-\varepsilon)(\rho_p - \rho_f)/\rho_p}$ $u_K = u_t n(1-\varepsilon)\varepsilon^{n-1}$
Conclusions:	If $u_K > u_D$, the bed starts to bubble at the minimum fluidization condition: $\varepsilon_{mb} = \varepsilon_{mf}$ (=0.4). If $u_K < u_D$, the bed is initially homogeneous. To find ε_{mb}, progressively increase ε, repeating step 5 until $u_K = u_D$: $\varepsilon = \varepsilon_{mb}$. If condition $u_K < u_D$ persists over full expansion range, $1 > \varepsilon > 0.4$, then the bed is always homogeneous.

The homogeneous expansion region for gas fluidization of fine powders

In Chapters 9 and 12, the predictions of eqn (8.36) will be compared with the copious body of observations reported for the stability condition of a wide variety of experimentally investigated fluidized beds. By way of introduction to these comparisons, an example of the *particle bed model* predictions of the effect of particle diameter on the stability of gas fluidized beds – reflecting the observations reported at the start of this chapter, which confirmed the existence of the stable, homogeneously-fluidized state – is illustrated in Figure 8.3.

It will be seen that for the larger (150 μm) particles the 'stability limit' (where the dynamic and kinematic wave velocities intersect) occurs at a physically unobtainable void fraction (off scale in Figure 8.3), smaller than the packed bed value of 0.4. This indicates a system predicted to start bubbling ($u_K > u_D$) right from the minimum fluidization condition – behaviour typical for 'normal' gas fluidization. For the smaller 70 μm, particles the

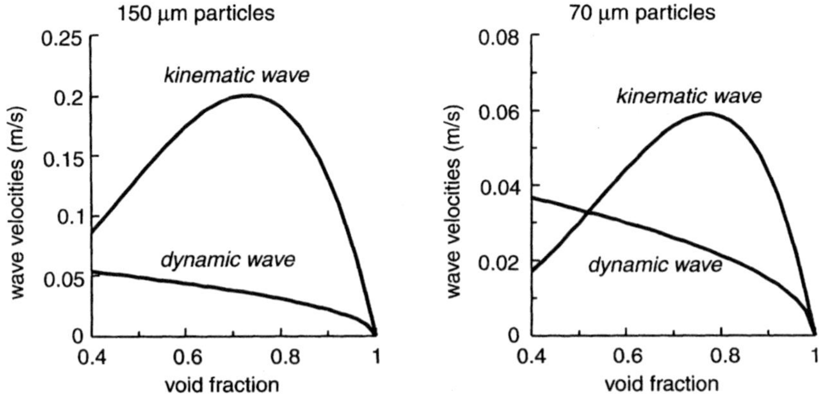

Figure 8.3 Dynamic- and kinematic-wave velocities as functions of void fraction for the fluidization of alumina particles by air ($\rho_p = 1000\,\text{kg/m}^3$, $d_p = 150$ and $70\,\mu\text{m}$).

model predicts the by now well-established condition of an initial region of homogeneous fluidization ($u_D > u_K$), extending to the critical, minimum bubbling point ($\varepsilon_{mb} = 5.2$), followed by bubbling behaviour.

The compressible fluid analogy

In Chapter 6, the stability criterion (derived formally above) was obtained by treating the particle phase of a fluidized suspension as analogous to a compressible fluid. In this way, the expression for the sonic velocity in an ideal gas, $\sqrt{\partial p/\partial \rho_f}$, was related to a concentration-gradient-induced voidage wave, resulting in the relation for the *dynamic-wave* velocity u_D. The *particle bed model*, eqns (8.21) and (8.26), may be readily cast in the form for a compressible fluid, thereby validating the sonic-velocity analogy.

The particle-phase density ρ_{pp} is defined in Chapter 6:

$$\rho_{pp} = (1 - \varepsilon)\rho_p. \tag{8.37}$$

This relates to a local mean value for a particle phase, which is regarded as occupying all the available volume – just as is understood by the term gas density: the solid particles are in this sense analogous to the molecules of a gas, and compression of the particle phase simply implies bringing the particles closer together.

Multiplying eqn (8.21) by ρ_p and applying the above definition for ρ_{pp} yields:

$$\frac{\partial \rho_{pp}}{\partial t} + \frac{\partial}{\partial z}(u_p \rho_{pp}) = 0. \quad \textit{Continuity} \tag{8.38}$$

The last term of the momentum equation (8.26) represents the 'particle pressure' gradient, $\partial p_p/\partial z$ (see eqns (8.15) and (8.19)). On this basis, eqn (8.26) becomes:

$$\rho_{pp}\left[\frac{\partial u_p}{\partial t} + u_p \frac{\partial u_p}{\partial z}\right] = F + \frac{\partial p_p}{\partial z}, \quad \textit{Momentum} \tag{8.39}$$

where the constitutive expression for F may be expressed as a function of u_p and ρ_{pp}.

Eqns (8.38) and (8.39) are in the form of the equations of change for a compressible fluid, of density ρ_{pp}, for which the sonic velocity under adiabatic conditions is given by $\sqrt{\partial p_p/\partial \rho_{pp}}$.

References

Foscolo, P.U. and Gibilaro, L.G. (1987). Fluid-dynamic stability of fluidized suspensions: the particle bed model. *Chem. Eng. Sci.*, **42**, 1489.

Geldart, D. (1973). Types of gas fluidization. *Powder Technol.*, **7**, 285.

Wallis, G.B. (1962). One-dimensional waves in two-component flow (with particular reference to the stability of fluidized beds). United Kingdom Atomic Energy Authority. Report AEEW-R162.

Wallis, G.B. (1969). *One-Dimensional Two-Phase Flow*. McGraw-Hill.

9

Single-phase model predictions and experimental observations

Powder classification for fluidization by a specified fluid

The *particle bed* model described in the previous chapter makes quantitative predictions concerning the stability of the homogeneously fluidized state – a feature that enables it to be rigorously tested. Experimental observations of fluidized-bed stability are reported extensively in the literature for a wide variety of systems. The copious data for the void fraction ε_{mb} at the minimum bubbling point, for gas-fluidized beds that exhibit a transition from homogeneous to bubbling fluidization, are compared directly with the single-phase model predictions in the following section. Before that, however, it will be demonstrated how the stability criterion may be used to construct global maps for the general classification of powders with regard to the type of fluidization (homogeneous, bubbling, or a homogeneous–bubbling transition) predicted for *any specified fluid*, thereby providing compact means for assessing large classes of data.

The criterion of eqn (8.36) was previously reported in a dimensionless form, eqn (6.10), which defines the *stability function S*:

$$S = (u_D - u_K)/u_k$$

$$= \frac{1.79}{n} \cdot \left(\frac{gd_p}{u_t^2}\right)^{0.5} \left(\frac{\rho_p - \rho_f}{\rho_p}\right)^{0.5} \left(\frac{\varepsilon^{1-n}}{(1-\varepsilon)^{0.5}}\right) - 1 = \begin{array}{rl} +\text{ve}: & \text{homogeneos} \\ 0: & \text{stability limit} \\ -\text{ve}: & \text{bubbling} \end{array} \qquad (9.1)$$

For a *particular fluid* at a given temperature and pressure, only the particle parameters, ρ_p and d_p, remain to be specified in S in order for the stability condition of the system to be predicted over the full working range of void fraction, $1 > \varepsilon > 0.4$. It is a simple matter, as will now be demonstrated, to use selected values of ρ_p and d_p to construct a predictive *stability map*, Figure 9.1, for the fluidization of any powder by the chosen fluid. This diagram shows at a glance how a given powder will fluidize: always homogeneously, always in the bubbling mode, or with a transition from homogeneous to bubbling behaviour.

The curves shown in Figure 9.2 indicate the various possibilities that S can display as a function of void fraction ε. For a given particle density and a specified fluid, the effect of increasing the particle diameter (the only remaining parameter) is to lower the minimum of S, thereby tending

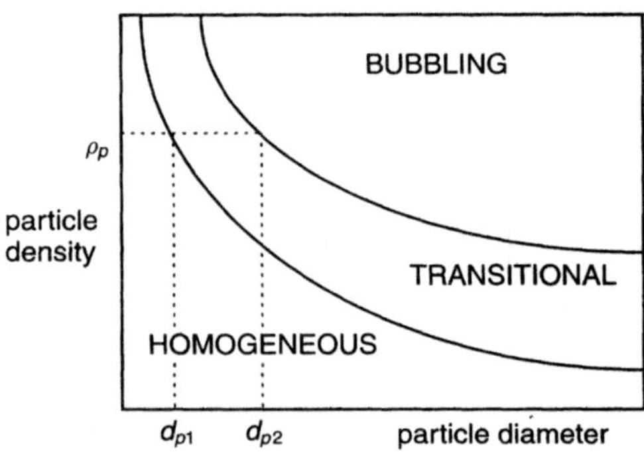

Figure 9.1 Stability map for a specified fluid.

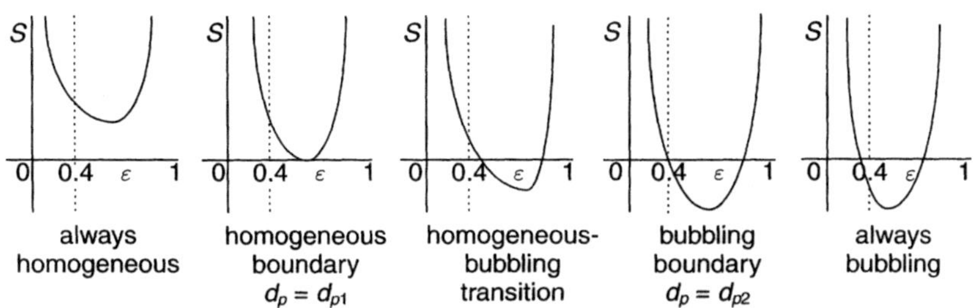

Figure 9.2 The stability function.

to change the behaviour progressively rightwards through the sequence illustrated in Figure 9.2.

The far left-hand curve of Figure 9.2 shows S to be always positive, so that there is no solution for the stability limit, $S = 0$: this represents a system that always fluidizes homogeneously. Increasing the particle diameter tends to shift the curve downwards; the particle diameter d_{p1} which causes the curve to just touch the ε-axis (second curve from the left) fixes a point on the *homogeneous-transitional* boundary of the stability map, as shown in Figure 9.1; the full boundary is obtained by finding this critical particle diameter for a range of values of particle density.

The third curve of Figure 9.2 shows a first stability limit, $S = 0$, at a void fraction ε_{mb} greater than ε_{mf} (shown as 0.4), indicating a system that exhibits a transition from homogeneous to bubbling behaviour; in these cases there is always a second transition back to the homogeneous state, which generally occurs close to the upper expansion limit, $\varepsilon \to 1$; we shall see in Chapter 12 that, for some liquid systems, this second transition point can be attained in practice.

The fourth curve shows a stability limit at the minimum fluidization point, $\varepsilon_{mb} = \varepsilon_{mf} = 0.4$; the particle diameter d_{p2} at which this occurs defines a point on the *bubbling-transitional* boundary of the stability map as shown in Figure 9.1; the full boundary is obtained, as before, by finding this second critical point for varying values of particle density.

Larger particle systems are predicted always to exhibit bubbling behaviour – typical of 'normal' gas fluidization; this case is represented in the far right-hand curve of Figure 9.2, which shows a physically unrealizable prediction for ε_{mb} of less than 0.4: such systems are unstable right from the minimum fluidization condition.

Fluidization by ambient air

The global map for fluidization by ambient air, constructed as described above, is shown in Figure 9.3. The right-hand boundary, which separates powders that always fluidize in the bubbling mode from those that exhibit a transition from the homogeneous to the bubbling mode (at the critical fluid flux U_{mb} and void fraction ε_{mb}) is well documented on the basis of experimental observations: the broken line represents this boundary in the Geldart (1973) empirical powder classification, in reasonable agreement with the predictive relation.

The left-hand boundary shown in Figure 9.3 is of less immediate physical significance, as the very fine powders that it largely represents (having diameters generally less than about 20 μm) do not usually fluidize well – if at all – due to the influence of adhesive contact forces that tend to stick them together in the packed state. Powders to the left of this boundary that do fluidize, do so homogeneously as predicted. An interesting example of this phenomenon is described by Akapo (1989); very fine silica hydrogel particles ($d_p < 20$ μm) were found to be initially cohesive, but to fluidize homogeneously after the surface forces had been neutralized by chemical treatment. And in general, the boundary itself

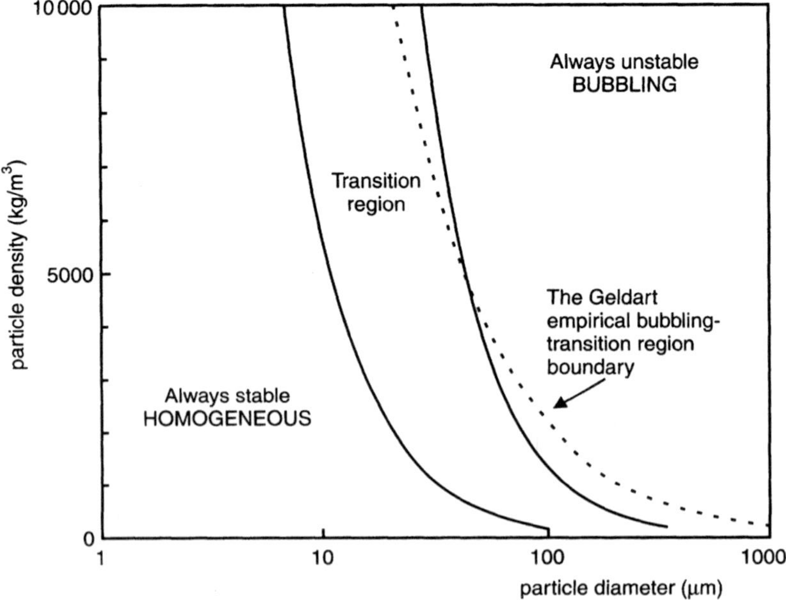

Figure 9.3 Stability map for fluidization by ambient air.

bears some relation to that separating the homogeneous-bubbling transition region from the cohesive, often unfluidizable, region of the Geldart (1973) classification. A tentative explanation for this convenient convergence has been attempted (Gibilaro *et al.*, 1988) and is discussed in Chapter 10, which also displays a more complete comparison of *particle bed* model predictions and the Geldart classification.

The minimum bubbling point

In this section, direct comparisons of the model predictions are made with reported experimental observations for gas-fluidized beds that display a transition from homogeneous to bubbling behaviour at the critical void fraction ε_{mb}. Many experimental data points are available for this purpose. As an aid to the evaluation of the comparisons that follow, we start with a brief discussion of some of the major sources of experimental error and of the sensitivity to error of the model predictions.

Sources of experimental error

Premature bubbling

It is important to bear in mind that the predictions of ε_{mb} arise from a *linear* analysis, and so relate only to *small* perturbations about the equilibrium state. In conducting experiments to measure ε_{mb} it is therefore essential to take precautions to avoid equipment-induced disturbances that exceed the linear response limit of the system. Major disturbances can result from inefficient fluid distribution, so it is important to provide fluid stabilization before the distributor, and a sufficient pressure drop across it. Any bed internals that disrupt the flow path, such as thermometer pockets and heat-exchanger tubes, should be removed, and sources of mechanical vibration should be neutralized.

It has long been known that such disturbance sources can give rise to the phenomenon of *premature bubbling*: that is to say a measured ε_{mb} value lower than that obtained in a disruption-free system. For this reason the homogeneous expansion behaviour of fine-powder, gas-fluidized beds has sometimes been referred to as *metastable* – because the stability can be destroyed by simply disrupting the regular flow operation. In Chapter 14, an analysis that employs the full, non-linear, *particle bed* model formulation provides a simple explanation of this phenomenon.

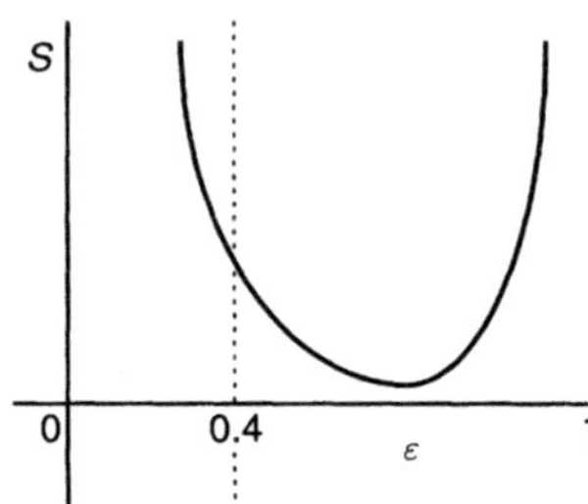

Figure 9.4 Model sensitivity case.

Model sensitivity

There can be a problem with the sensitivity of the model predictions in cases where the system under examination lies close to the boundary between the homogeneous and transitional regions on the stability map. Fortunately, most of the reported results for ε_{mb} relate to gas fluidized, fine-powder systems well away from this boundary, so that severe sensitivity problems do not arise; on the occasions when they do, however, it is as well to be aware of them.

For problem cases, the stability function S varies with ε as shown in Figure 9.4; the minimum is close to the ε axis, either just above or just below it. Under these circumstances a very small error in the evaluation of S can have the effect of changing the prediction from an ε_{mb} value of, typically, somewhere around 0.7 (where the curve first cuts the ε axis) to the 'always homogeneous' condition shown in Figure 9.4. This means that predictions of a continuous, gradual increase in ε_{mb} (as a result, say, of a progressive decrease in particle diameter) can suddenly be followed, at a ε_{mb} value well short of unity, by a jump to the 'always homogeneous' prediction.

Gas fluidization

We now present results, reported by various workers, for experimental ε_{mb} determinations. Particle property variations, ρ_p and d_p, involve changing the bed inventory, whereas the fluid properties, ρ_f and μ_f, can be varied continuously by simply altering the operating pressure and temperature. We start with examples of reported observations in which a single parameter (ρ_p, d_p, ρ_f, μ_f) is varied systematically. These are the most useful experiments for comparative purposes; the trends uncovered are more informative and less susceptible to error than the absolute values themselves. Further such examples are reported by Gibilaro *et al.* (1988).

The effect of gas pressure (gas density)

The progressive increase in the minimum bubbling point with increasing pressure for fine-powder gas fluidization was first reported by Rowe *et al.* (1984); it was subsequently investigated over a much greater range of pressure by Jacob and Weimer (1987). Extensive regions of homogeneous behaviour were observed. The system consisted of granular carbon particles ($d_p = 44$ and $112\,\mu m$) fluidized by synthesis gas at ambient temperature. The pressure was varied from 20 to 120 bar. At each selected pressure, the experimental homogeneous expansion characteristics (ε as a function of gas flux U_0) were used to determine u_t and n; these evaluations (rather than those that may be obtained from the general correlations) are used for the ε_{mb} values that are compared with measured ones in Figure 9.5. (At 120 bar, the 44 µm particles are predicted to always fluidize homogeneously – reflecting the 'model sensitivity' issue addressed above.)

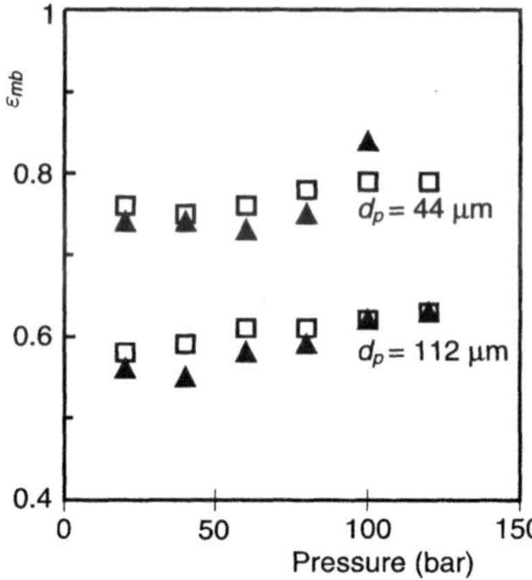

Figure 9.5 Effect of gas density on the minimum bubbling point: comparison of the results of Jacob and Weimer, 1987 (open squares) with model predictions (solid triangles).

The effect of temperature (gas viscosity)

The predominant effect on the system parameters of an increase in gas-fluidization temperature is an increase in gas viscosity. The effect of temperature on the minimum bubbling point for three catalyst particle beds fluidized by nitrogen has been reported by Rapagnà *et al.* (1994): the temperature range extended from ambient to nearly 1000 °C. Two of these systems exhibited homogeneous to bubbling transitions, which

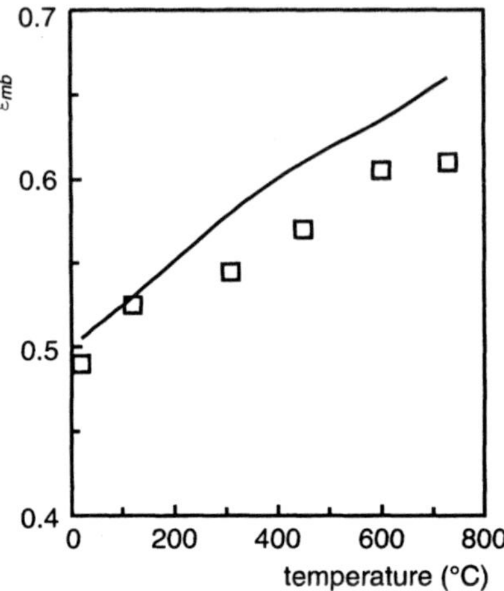

Figure 9.6 Effect of gas viscosity on the minimum bubbling point: comparison of the results of Rapagnà *et al.* (1994) with model predictions.

varied with temperature in good agreement with the predictions of the stability criterion: one of these (equivalent diameter $d_p = 103\,\mu m$, $\rho_p = 1500\,kg/m^3$) is illustrated in Figure 9.6, in which the experimental results are shown as points, and the model predictions (using the u_t and n correlations of Table 8.1) as a continuous curve.

The effect of particle size

A systematic study of the influence of particle size on the minimum bubbling point for alumina catalyst powder ($\rho_p = 850\,kg/m^3$) fluidized by ambient air was reported by De Jong and Nomden (1974). Their results for ε_{mb} as a function of d_p are shown as points in Figure 9.7. The predictions of the stability criterion, eqn (9.1), obtained using the u_t and n correlations of Table 8.1, are shown as a continuous curve. Note that ε_{mb} predictions of less than 0.4 translate, for a physical system, into bubbling behaviour right from the onset of fluidization; hence the constant predicted ε_{mb} value of 0.4 for particles somewhat larger than $100\,\mu m$.

The effect of particle density

A systematic study, such as the one reported above for particle size variation, does not appear to be available for the effect of particle density on the minimum bubbling point. However, by gathering together

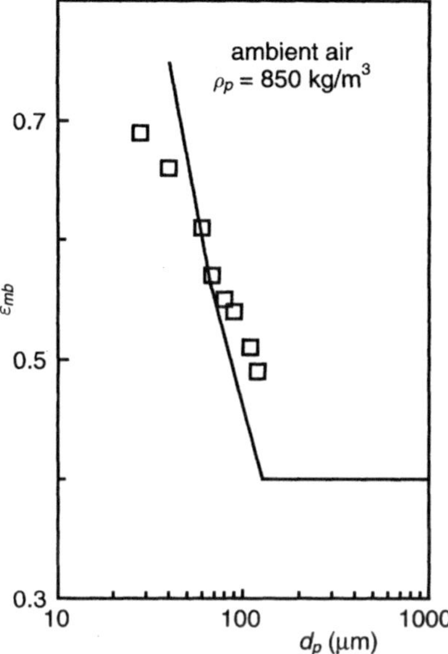

Figure 9.7 Effect of particle size on the minimum bubbling point: comparison of the results of De Jong and Nomden (1974) with model predictions.

the published data for ambient air and nitrogen fluidization of various materials, a clear picture of this dependency emerges. Figure 9.8 shows reported ε_{mb} values as data points for particles of approximately 60 μm; also shown, as a continuous curve, are the predictions of the criterion, eqn (9.1). The 10 data points are from 10 separate studies, referenced in Gibilaro *et al.* (1988), which also presents similar comparisons for 40 μm and 100 μm particles.

The effect of gravitational field strength

The *particle bed* model readily accommodates variations in *g*. When the comparisons with model predictions that we now reproduce (Gibilaro *et al.*, 1986) and the experiments to which they refer were first performed the whole exercise appeared purely academic. It was subsequently gratifying to learn that a space-exploration study had given serious consideration to the feasibility of operating fluidized bed reactors under conditions of greatly reduced gravitational field strength. Without an effective model, it is quite impossible to predict what effect such an environment would have on the overall performance.

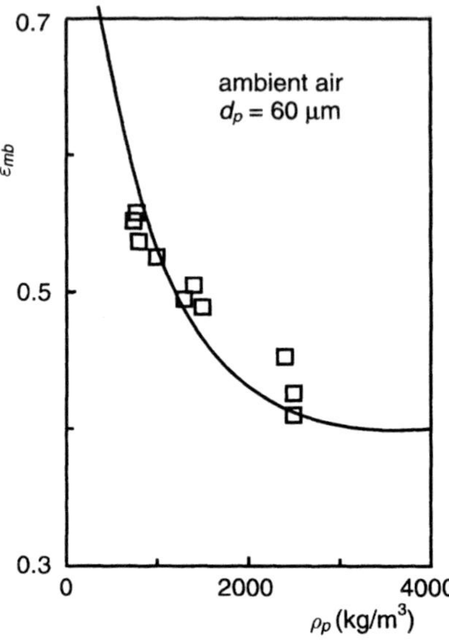

Figure 9.8 Effect of particle density on the minimum bubbling point: comparison of the results of 10 separate studies (see Gibilaro *et al.*, 1988) with model predictions.

Although experimental results do not appear to be available for low g, an ambitious programme for determining the effect on bed stability of simulated high gravitational field strength g_s has been reported by Rietema and Mutsers (1978). This involved a 'human centrifuge', normally used for pilot training, fitted for the purpose in hand with a fluidized bed and ancillary equipment, which included a chemical engineering PhD student. Experiments were performed on beds of cracking catalyst ($\rho_p = 1414\,\text{kg/m}^3$, $d_p = 62\,\mu\text{m}$) and polypropylene particles ($\rho_p = 920\,\text{kg/m}^3$, $d_p = 40\,\mu\text{m}$), each fluidized by both nitrogen and hydrogen. Effective gravitational field strengths g_s of up to three times the ambient level g were attained. Measurements were made of the minimum bubbling points, which were reported as functions of g_s/g for each bed.

The *particle bed* model predicts a decrease in stability with increasing gravitational field strength, and this trend was confirmed in every case. For the nitrogen-fluidized systems (shown in Figure 9.9) the absolute agreement was also excellent, especially for the cracking catalyst beds, whereas for hydrogen-fluidization the ε_{mb} predictions were consistently some 0.1 below the reported levels.

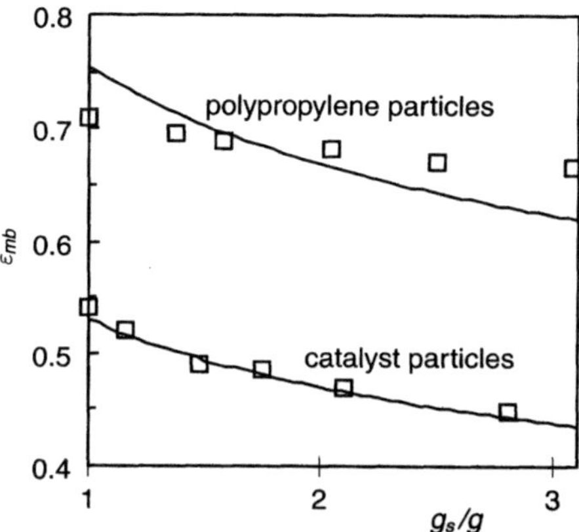

Figure 9.9 Nitrogen fluidization at high simulated gravitational field strength: comparison of the results of Rietema and Mutsers (1978) (points) with model predictions (continuous curves).

Other reported ε_{mb} measurements in gas-fluidized beds

All the ε_{mb} data points discussed above, together with other available published values, have been used to compile Figure 9.10. References to 29 literature sources from which they have been drawn are listed by Foscolo *et al.* (1995).

The data points shown, for want of a more appropriate location, at the extreme right of Figure 9.10 (on the ε_{mb} (predicted) $= 1$ boundary) represent systems predicted to remain always homogeneous but which in fact exhibited a switch to bubbling behaviour. A likely cause for these discrepancies is the *premature bubbling* phenomenon discussed above, resulting from flow disturbances caused by such things as inadequate fluid distribution at the entry region, or physical obstructions to the flow pathways. The lack of continuity of these points with the main body of data is a consequence of the *model sensitivity* phenomenon, also discussed above; for systems in which the progressive variation of a parameter leads to a progressive increase in stability, there exists an upper limit for ε_{mb} at a value well below unity, corresponding to the second curve of Figure 9.2. Another feature of Figure 9.10 that is worthy of note is the fact that no *experimental* data exists for ε_{mb} greater than about 0.8, in spite of the existence of systems predicted to always fluidize homogeneously. This second lack of continuity, this time with regard to experimental observations, is fully in keeping with the predicted form of the stability function S (Figure 9.2), and therefore well in accord with the model predictions.

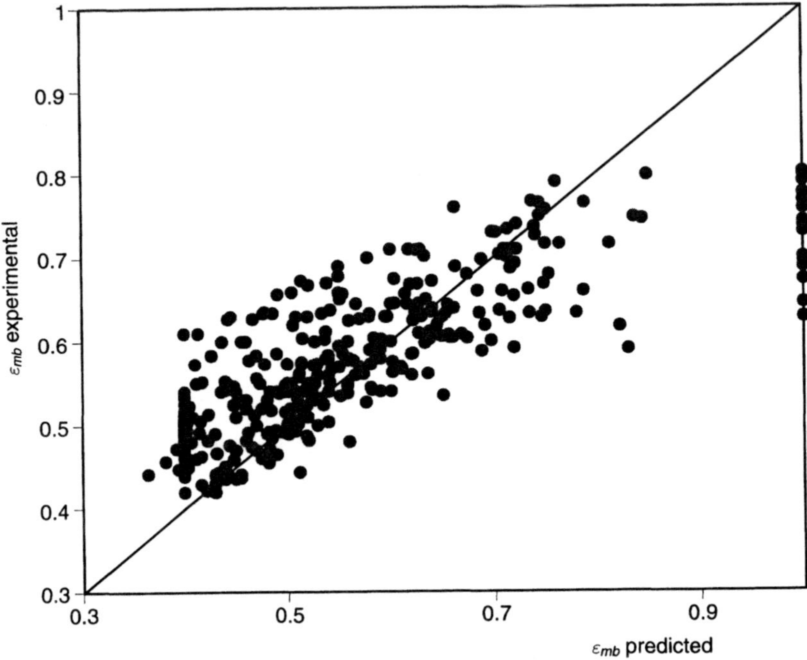

Figure 9.10 Comparison of available reported ε_{mb} data points with model predictions.

The wave velocities

The extensive experimental evidence for the predictive ability of the stability criterion, summarized in the preceding sections, provides indirect support for the constitutive relations employed for the kinematic- and dynamic-wave velocities. We now consider more direct means of measuring these quantities.

The kinematic-wave velocity u_K

In Chapter 5, we saw that homogeneous liquid-fluidized beds subjected to a sudden reduction in fluid flux respond in a very simple manner: the bed surface falls at constant velocity u_{bs} and a kinematic-shock travels up from the distributor at velocity u_{KS}:

$$u_{KS} = (1 - \varepsilon_1) \cdot \frac{u_{bs}}{\varepsilon_2 - \varepsilon_1}, \tag{9.2}$$

where ε_1 and ε_2 are the equilibrium void fractions of the bed before and after the fluid flux change. Equation (9.2) is simply eqn (5.8) written in terms of the bed surface velocity u_{bs} – a readily measurable quantity.

The kinematic-wave velocity u_K is the limiting value of u_{KS} as $\varepsilon_1 \to \varepsilon_2 \to \varepsilon$. It may therefore be evaluated by performing a number of bed collapse experiments of varying flux change magnitude, $\Delta U = U_2 - U_1$, that all correspond to the same *mean* void fraction ε; u_K is then obtained by measuring the bed surface collapse rate u_{bs} in each case, evaluating the kinematic-shock velocity u_{KS} as a function of ΔU by means of eqn (9.2), and extrapolating back to $\Delta U = 0$. Example applications of the method are illustrated in Figure 9.11.

Values of u_K for water fluidization of the copper and glass particles used in the above illustration have been obtained for various expansion conditions (Gibilaro *et al.*, 1989). In Figure 9.12 these results are compared with the theoretical expression:

$$u_K = nu_t(1 - \varepsilon)\varepsilon^{n-1}. \tag{9.3}$$

Water-fluidization experiments are discussed in more detail in Chapter 12, after the derivation of a stability criterion in Chapter 11 that is based on the full, two-phase model – which is more appropriate for liquid-fluidized systems for which particle and fluid densities are relatively close. It will be seen, however, that the kinematic-wave velocity expression emerging from this more complete description is identical to that of the simplified, one-phase treatment considered here, eqn (9.3).

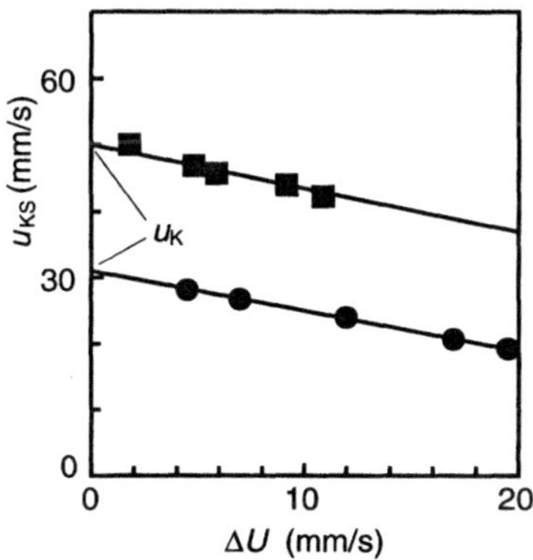

Figure 9.11 Experimental determination of u_K: fluidization of 275 μm copper (upper line, $\varepsilon = 0.55$) and 550 μm soda glass (lower line, $\varepsilon = 0.70$) by ambient water.

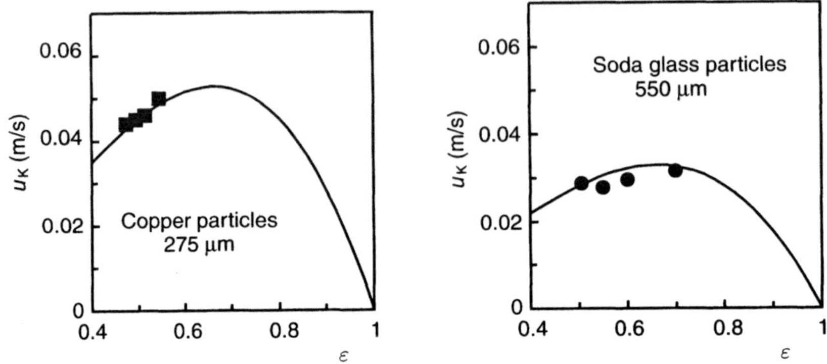

Figure 9.12 Kinematic-wave velocities in water-fluidized beds: comparison of experimental evaluations (points) with theoretical predictions (continuous curve, eqn (9.3)).

Large wavelength voidage waves

A further confirmation of the relation for u_K is to be found from reported measurements of high void fraction bands in 'two-dimensional' liquid-fluidized beds. These beds consist of two sheets of plane glass separated by a narrow gasket. They have been widely used to study fluidized-bed inhomogeneities, which are difficult to observe in normal, three-dimensional equipment. The first such studies (Hassett, 1961a, 1961b) reported a uniform, wave-like behaviour, consisting of high void fraction, upwards-propagating horizontal bands – Figure 9.13.

Careful measurements of the properties of these waves were subsequently made by El-Kaissy and Homsy (1976) for glass particles of four sizes, each fluidized at three different fluid fluxes. We see in Chapter 10

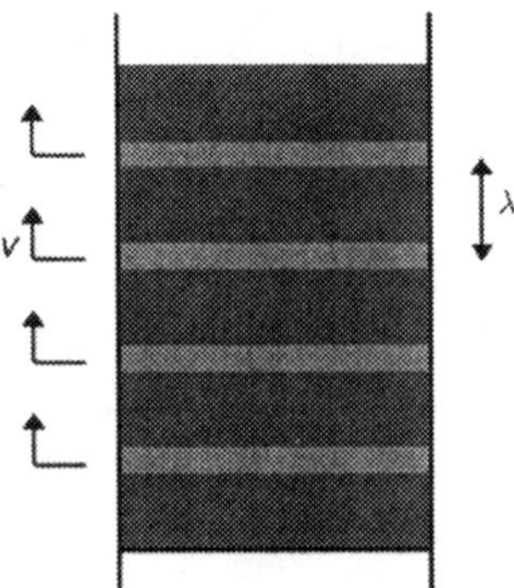

Figure 9.13 High void fraction bands in water-fluidized glass beds.

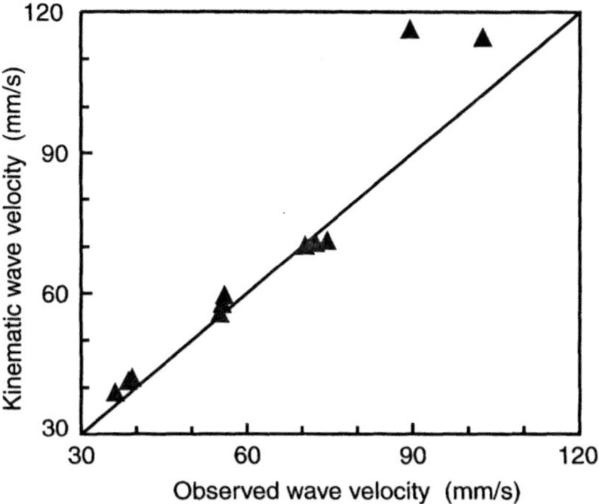

Figure 9.14 Wave propagation in water-fluidized beds: comparison of experimental measurements of El-Kaissy and Homsy (1976) with model predictions.

that the *particle bed* model leads to predictions of wave perturbation velocities v in fluidized beds: for long wavelengths λ (the case for all the 12 reported results) the waves are predicted to travel at the kinematic-wave velocity u_K. Figure 9.14 compares predicted values of u_K, eqn (9.3), with the perturbation wave velocities reported by El-Kaissy and Homsy (1976); on the whole agreement is very good, fully endorsing the kinematic-wave velocity relation, as well as the *particle bed* model predictions.

The dynamic-wave velocity u_D

The kinematic-wave velocity relation is well-established, and experimental evaluations of its validity, such as those described above, are easy to conceive and execute. This is by no means the case for the dynamic-wave velocity relation,

$$u_D = \sqrt{3.2gd_p(1-\varepsilon)(\rho_p - \rho_f)/\rho_p}, \qquad (9.4)$$

which is both controversial and difficult to validate directly by experiment. The experimental difficulties have been summarized by Wallis (1962), who also devised the partial resolution of the problem, which we now describe.

The difficulties concern the fact that in a stable system dynamic waves run into the slower kinematic waves, whereas in an unstable system they

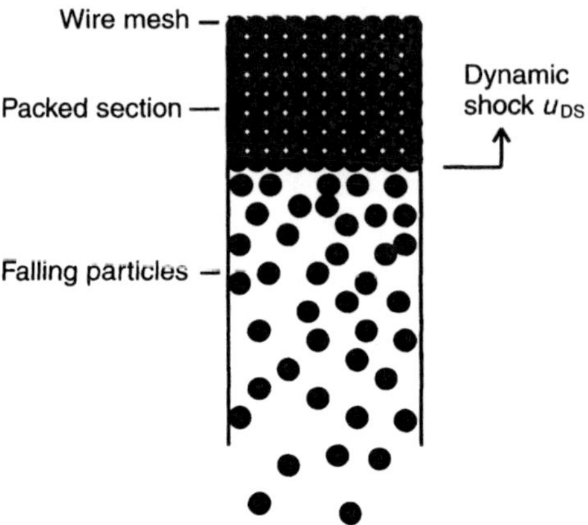

Figure 9.15 Experimental determination of dynamic-shock velocity.

grow rapidly in amplitude to develop into shocks (or bubbles). Some ingenuity is called for in designing appropriate experiments for dynamic-wave velocity determination. (There is some confusion on this matter resulting from reported measurements of gas compression-wave velocities in gas-fluidized beds, which are slowed down somewhat due to the presence of the particles. These easily-measured velocities bear no relation at all to those of the dynamic wave for the particle phase.) The experimental technique of Wallis (1962) for measuring u_D is illustrated in Figure 9.15.

The method involves fitting a mesh to the top of the bed tube, and then packing all the particles against it by increasing the fluid flux to a sufficiently high level. Whereas the minimum fluidization flux U_{mf} should be just sufficient to maintain the bulk of the particles in place once they have been compacted against the mesh, it has long been known (Rowe and Henwood, 1961) that, under these conditions, particles at the bottom interface become detached and start to 'rain down' at fluid fluxes a little below about $2U_{mf}$: this gives rise to an upwards-propagating 'particle phase expansion shockwave' (a *dynamic-shock* u_{DS}). We shall see in Chapter 14 that such expansion shocks are intrinsically unstable, and therefore not sustainable in a truly fluidized system. For the situation considered here, however, the presence of contact forces, transmitted from the upper mesh through the packed particle assembly, appears to stabilize the expansion shock, rendering its velocity easily measurable.

The experimental procedure thus entails: first packing all the particles against the upper mesh; then reducing the fluid-flux U to a value in the

Table 9.1 Comparison of measured dynamic-wave velocities with those predicted by eqn (9.4) under incipient fluidization conditions

	Particles		*Dynamic-wave Velocity* u_D	
Material	ρ_p (kg/m^3)	d_p (μm)	*Measured* (mm/s)	*Predicted* (mm/s)
Copper	8600	550	97	96
	8800	275	81	68
Zirconia	3800	1800	180	158
	3800	750	105	102
Lead glass	2900	1100	119	117
	2900	655	99	90
	2900	425	75	73
Soda glass	2500	2200	154	158
	2500	1100	112	112
Resin	1420	5000	95	167
	1270	4000	82	127
	1270	2000	41	89

range $2U_{mf} > U > U_{mf}$, and measuring the velocity u_{DS} of the resulting dynamic shock. By performing a number of such experiments at different U values, the *dynamic-wave* velocity u_D under incipient fluidization conditions (U_{mf}, ε_{mf}) may be obtained by extrapolation to $U = U_{mf}$. Results obtained in his way for water fluidization of various particle species (Gibilaro *et al.*, 1989) are reproduced in Table 9.1.

The above comparisons show very good agreement of measured with predicted u_D values, except for the three low-density resin particles, for which the measured values fall well short of the predictions. This mismatch may be attributed to two omissions in the model formulation adopted so far, both of which become increasingly significant as the particle density approaches that of the fluid. The first is a consequence of the assumption that particle density is much greater than fluid density, which effectively reduces the two-phase system to a single phase. We shall see in Chapter 12 that this assumption results in negligible error for all cases of gas fluidization, and only becomes of real significance for the liquid fluidization of particles of low density – such as the resins reported in Table 9.1. We here anticipate the results of the more complete (two-phase) formulation of the *particle bed* model, reported in Chapter 11, by

Table 9.2 Correction of predicted u_D for two-phase model and added mass effects

Resin particles		*Dynamic-wave velocity u_D* (mm/s)			
ρ_p (kg/m^3)	d_p (μm)	*Measured*	*One-phase model*	*Two-phase model*	*Two-phase model + added mass*
1420	5000	95	167	127	109
1270	4000	82	127	92	77
1270	2000	41	89	64	53

correcting the predictions for the resin particle systems for this factor. The other omitted effect concerns the phenomenon of *added mass*. This we now briefly consider, and report how it may be partially corrected for in predictions of fluidized-bed behaviour.

Added mass effects

The *particle bed* model formulation so far employed omits consideration of *inertial coupling* phenomena. These arise when a body submerged in a fluid is subjected to a net force causing it to accelerate; as it does so, some fluid is carried with the body, effectively increasing its inertial mass. Clearly, this *added mass* will be negligible for gas-fluidized systems, but for liquid fluidization of relatively low-density particles it could well be important. A remarkably simple method of approximating this effect for the problem in hand has been derived by Wallis (1990). His analysis leads to the conclusion that some correction can be obtained by simply increasing the value of both the particle and fluid densities in the model formulation (and hence in any derived result) by one-half the fluid density:

$$\rho_p \rightarrow \rho_p + \frac{\rho_f}{2}, \quad \rho_f \rightarrow \rho_f + \frac{\rho_f}{2}. \tag{9.5}$$

The results of both corrections to the dynamic-wave velocity predictions for the resin particle systems of Table 9.1 are shown in Table 9.2 (Gibilaro *et al.*, 1990).

Figure 9.16 compares all available dynamic-wave velocity predictions with measurements made under incipient fluidization conditions as described above. These consist of: those reported in Table 9.1 (solid

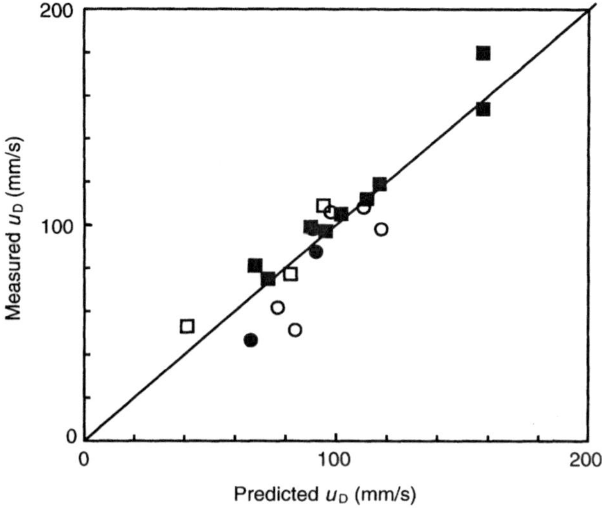

Figure 9.16 Comparison of measured and predicted dynamic-wave velocities under incipient fluidization conditions: $\varepsilon = \varepsilon_{mf} = 0.4$.

squares), with the low-density, resin-particle systems corrected for the fluid pressure field and added mass effects as reported in Table 9.2 (open squares); and the original results of Wallis (1962) for both air and water fluidization of various particles (open and solid circles, respectively).

The direct verification of the kinematic- and dynamic-wave speed expressions, which in some respects effectively define the *particle bed* model, provides further support for the formulation. Still required, however, is an experimental method for measuring dynamic-wave speeds in the particle phase of expanded beds, at void fractions greater than ε_{mf}.

Conclusions

Taken as a whole, the results reported in this chapter provide overwhelming support for the essential mechanistic integrity of the model formulation. The stability criterion itself has been shown to differentiate between typical liquid and gas fluidization (respectively homogeneous and bubbling); between fine-powder gas systems that exhibit an initial region of homogeneous expansion, and those that bubble from the onset of fluidization (illustrated in the global map for fluidization by ambient air); and even to provide reasonable estimates of the precise point at which homogeneous gas fluidization gives way to bubbling behaviour.

Perhaps the most significant feature of these comparisons is the way in which the trend in predicted ε_{mb} values has been shown to be faithfully reproduced in experimental investigations involving the systematic variation of every variable (ρ_p, d_p, ρ_f, μ_f and even g) that enters into the *fluid-dynamic* description of the fluidization process. This represents crucial evidence in the controversy that still exists in certain quarters regarding the general mechanism responsible for a homogeneous gas-fluidized state. For historical reasons, illuminated by the analysis reported in Chapter 7, the fluidized state came to be regarded as intrinsically unstable from a purely fluid-dynamic point of view. This has had the effect of directing attention to possible *non-fluid-dynamic* interventions in the search for an explanation for observed stability. These involved particle–particle interactions, attributable to such things as electrostatic, capillary, van der Waals's and direct collision forces. Although situations certainly exist in which direct particle–particle interactions become important (for example, in the fluidization of electrically-charged resin particles, or ferromagnetic particles fluidized in the presence of an electromagnetic field), such systems are characterized by strongly anomalous behaviour, quite out of line with general empirical trends, such as those embodied in the Geldart powder classification map.

References

Akapo, S.O. (1989). *Gas–solid fluidization: an improved method for the preparation of chemically bonded stationary phases*. PhD Thesis, University of London.

De Jong, J.A.H. and Nomden, J.F. (1974). Homogeneous gas–solid fluidization. *Powder Technol.*, **9**, 91.

El-Kaissy, M.M. and Homsy, G.M. (1976). Instability waves and the origin of bubbles in fluidized beds. *Int. J. Multiphase Flow*, **2**, 379.

Foscolo, P.U., Gibilaro, L.G. and Rapagnà, S. (1995). Infinitesimal and finite voidage perturbations in the compressible particle-phase description of a fluidized bed. In: Developments in Fluidization and Fluid–Particle Systems (J.C. Chem, ed.). *AIChE Symposium Series*, **91**(308), 44.

Geldart, D. (1973). Types of fluidization. *Powder Technol.*, **7**, 275.

Gibilaro, L.G., Di Felice, R. and Foscolo, P.U. (1986). The influence of gravity on the stability of fluidized beds. *Chem. Eng. Sci.*, **41**, 2438.

Gibilaro, L.G., Di Felice, R. and Foscolo, P.U. (1988). On the minimum bubbling voidage and the Geldart classification for gas-fluidized beds. *Powder Technol.*, **56**, 21.

Gibilaro, L.G., Di Felice, R. and Foscolo, P.U. (1989). The experimental determination of one-dimensional wave velocities in liquid-fluidized beds. *Chem. Eng. Sci.*, **44**, 101.

Gibilaro, L.G., Di Felice, R. and Foscolo, P.U. (1990). Added mass effects in fluidized beds: application of the Geurst–Wallis analysis of inertial coupling in two-phase flow. *Chem. Eng. Sci.*, **45**, 1561.

Hassett, N.L. (1961a). Flow patterns in particle beds. *Nature*, **189**, 997.

Hassett, N.L. (1961b). The mechanism of fluidization. *Br. Chem. Eng.*, **11**, 777.

Jacob, K.V. and Weimer, A.W. (1987). High-pressure particulate expansion and minimum bubbling of fine carbon powders. *AIChE J.*, **33**, 1698.

Rapagnà, S., Foscolo, P.U. and Gibilaro, L.G. (1994). The influence of temperature on the quality of gas fluidization. *Int. J. Multiphase Flow*, **20**, 305.

Rietema, K. and Mutsers, S.M.P. (1978). The effect of gravity upon the stability of a homogeneously fluidized bed, investigated in a centrifugal field. *Fluidization*. Cambridge University Press.

Rowe, P.N. and Henwood, G.A. (1961). Drag forces in a hydraulic model of a fluidized bed. Part 1. *Trans. Inst. Chem. Eng.*, **39**, 43.

Rowe, P.N., Foscolo, P.U., Hoffman, A.C. and Yates, J.G. (1984). X-ray observations of gas-fluidized beds under pressure. In: *Fluidization 4* (D. Kunii and R. Toei, eds). Engineering Foundation.

Wallis, G.B. (1962). One-dimensional waves in two-component flow (with particular reference to the stability of fluidized beds). United Kingdom Atomic Energy Authority Report: AEEW-R162.

Wallis, G.B. (1990). On Geurst's equations for inertial coupling in two-phase flow. In: *Two-Phase Flow and Waves* (D.D. Joseph and D.G. Schaeffer, eds). Springer-Verlag.

10

Fluidization quality

Behaviour spectra for fluidization

In the preceding chapters we first considered the primary forces acting on a fluidized particle in a bed in equilibrium, and then the elastic forces between particles that come into play under non-equilibrium conditions. These two effects provide closure for the *particle bed* model, formulated in terms of the particle- and fluid-phase conservation equations for mass and momentum. Up to now, applications have focused on the stability of the state of homogeneous particle suspension, in particular for gas-fluidized systems for which the condition that particle density is much greater than fluid density enables the particle-phase equations to be decoupled and treated independently. The analysis has involved solely the linearized forms of these equations, and has led to a stability criterion that broadly characterizes fluidized systems according to three manifestations of the fluidized state: always stable – the usual case for liquids; always unstable – the usual case for gases; and transitional behaviour – involving a switch, at a critical fluid flux, from the stable to the unstable condition. This characterization has been achieved by evaluation of the minimum-bubbling void fraction ε_{mb}, a system property that thereby provides an immediate first measure of *fluidization quality*.

The term 'fluidization quality' may be applied to describe the various fluid-dynamic conditions brought about by the fluidization process itself. The optimal combination of these conditions depends on the particular application in hand. Thus if the bed is to be employed as a filter, a *stable*, homogeneous particle suspension is desirable. (De Luca *et al.* (1994) have studied the application of liquid-fluidized beds as filters for macromolecules in a biochemical broth.) The conversion in a fluidized reactor, on the other hand, depends on the extent of heat and mass transfer within and between the phases, and is thus strongly influenced by the mixing action of bubbles and other inhomogeneities associated with the *unstable* fluidized state; also on the fluid and particle flow characteristics that reflect such things as bubble size and frequency, fluid residence time distribution, phase holdups, and many other basic features that together determine the fluidization quality. The uncertainty inherent in respect of these factors poses real problems in process design. Unless a proposed system can be assumed to relate closely to one for which the fluid-dynamic characteristics are known, some method for predicting fluidization quality must be devised in order to quantify the process model by means of which conversion in an envisaged commercial unit is to be estimated.

One method for obtaining the necessary information is by means of the scaling relations for fluidization, which are discussed in Chapter 13. These relations enable laboratory experiments to be performed on relatively simple 'cold models' that relate, with regard to fluidization quality, to a proposed reactor. With the present state of the art, this procedure, although costly and time-consuming, remains the most dependable on offer. Less arduous alternatives to scaling experiments would require the development of a reliable model for fluidization dynamics, which could be solved numerically using boundary conditions that relate to the geometric features of a proposed design. A preliminary application of the *particle bed* model (extended to multi-dimensional form) to this goal is described in Chapter 16; although the work is still in its infancy, initial results give grounds for confidence in the development of practical applications in the not too distant future.

In this chapter we take a more detailed look at the wave solutions to the linearized model equations, which delivered the stability criterion. These, it will be seen, lead to predictive criteria for fluidization quality, relating it to measures of the extent of system instability to small, imposed void fraction perturbations.

A first measure of fluidization quality: the minimum bubbling void fraction ε_{mb}

We now examine the relation of *predicted* ε_{mb} values to fluidization quality. This turns out to provide considerably more insight into bed behaviour than has hitherto been appreciated, generalizing reported conclusions concerning the influence of *measured* ε_{mb} determinations over limited regions of applicability. To illustrate this relation we first consider the empirical Geldart (1973) powder classification, which has been briefly referred to in the previous chapter.

The Geldart empirical powder-classification map for fluidization by ambient air

This map subdivides fluidization quality into four broad categories, Groups A, B, C and D, according to the basic powder properties: particle diameter d_p (or the *surface/volume* average diameter, eqn (3.8), in cases of significant size distribution) and particle density ρ_p.

Group A powders (typically in the size range 30–100 μm) are those that exhibit a transition from homogeneous to bubbling behaviour at a gas flux in excess of the minimum fluidization value: $U_{mb} > U_{mf}$, $\varepsilon_{mb} > \varepsilon_{mf}$. The bubbles produced remain relatively small for these systems and, as a result of homogeneous expansion prior to the minimum bubbling condition, the dense phase of the bubbling bed is likely to remain 'aerated' at a void fraction in excess of ε_{mf} and somewhat below ε_{mb} (a phenomena discussed, following a non-linear analysis of *particle bed* model predictions, in Chapter 14). These features provide good conditions for most fluidized bed applications, improving progressively with increasing ε_{mb} (Kwauk, 1992).

Group B powders are larger (up to about 1 mm, depending on particle density) and fluidize in a progressively more unstable manner as both the particle size and density are increased. The dense phase remains at the minimum fluidization value ε_{mf}, and the bubbles, which rise faster than the interstitial gas, are large and grow rapidly, mainly as a result of coalescence, as they rise through the bed. Although the mixing action of the bubbles is far more intense than for Group A powders, this positive aspect is usually more than outweighed by the reduced gas–solid contact resulting from the increase in bubble size, and by problems of particle attrition and the subsequent elutriation of fines; also by the damage done by erosion to heat exchanger tubes and other bed internals.

Group C powders are the smallest (typically less than 30 μm). In the packed state there is often a tendency for these particles to stick together, making them difficult to fluidize; this stickiness can be attributed to such things as liquid-bridging capillary forces (for damp powders) and van der Waals's effects, etc. The result can be that the gas cuts channels through the bed rather than being distributed evenly around the particles, giving rise to pressure losses in the emerging gas that are insufficient for the suspension of the entire bed. Various methods have been used to facilitate the fluidization of these powders, including mechanical agitation, vibration, and chemical treatment to neutralize the surface forces.

Group D powders are the largest (in excess of about 1 mm), are generally of lower density (typically seeds, such as wheat, etc.), and are difficult to

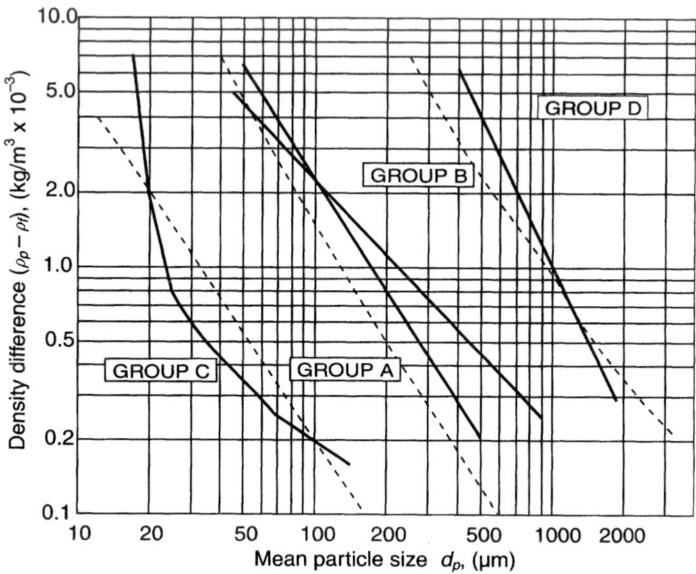

Figure 10.1 Powder classification map for fluidization by ambient air. Heavy lines, empirically determined boundaries of Geldart; light broken lines, boundary predictions of the *particle bed* model.

fluidize evenly. When fluidized, the dense phase remains at ε_{mf}, and the bubbles travel slower than the interstitial gas; particle mixing is poor.

The empirical boundaries separating these groups are shown as heavy lines in Figure 10.1. It should be emphasized that these lines provide only a rough guide, there being a good deal of data scatter, particularly with regard to the A/C and, to a lesser extent, the B/D boundaries. This is certainly to be expected for the A/C boundary, as the 'stickiness' responsible for Group C behaviour is of non-fluid dynamic origin and therefore hardly likely to be effectively correlated in terms of particle size and density. Changes in behaviour across the B/D boundary are by no means sharply defined, leading to a fair degree of uncertainty in its position. Even the A/B boundary can be affected by experimental error, particularly the 'premature bubbling' phenomenon discussed in Chapter 9 and analysed in Chapter 14.

A *predictive* powder classification map for fluidization by ambient air

Also shown in Figure 10.1, by means of light, broken lines, are *predictions* of the group boundaries obtained from the *particle bed* model. Only the middle one, the *A/B boundary*, is unambiguously defined, representing as it does the direct theoretical counterpart of the corresponding empirical A/B boundary in the Geldart map: it is simply the locus of (d_p, ρ_p) combinations that give rise to a void fraction at minimum bubbling ε_{mb} exactly equal to that at minimum fluidization ε_{mf} (≈ 0.4) – as described in the opening section of Chapter 9.

The construction of the derived *A/C boundary* is also described in the opening section of Chapter 9; it represents the dividing line between 'always homogeneous' and 'transitional' systems. The fact that this boundary should show some correspondence with that between the fine, 'cohesive' Group C powders and those that switch to bubbling behaviour above a certain critical void fraction might appear fortuitous, but may be tentatively attributed to the agitation induced by bubbles in the unstable fluidization regime at gas fluxes higher than U_{mb} (Gibilaro *et al.*, 1988); this action effectively replicates the mechanical agitation sometimes employed in the fluidization of Group C powders. Systems to the left of this boundary, which only fluidize homogeneously, possess no inherent disruptive mechanism with which to overcome the cohesive contact forces between particles. The fact that the proposed theoretical boundary is obtained on the basis of fluid-dynamic considerations is consistent with Geldart's correlation in terms of solely fluid-dynamic variables, ρ_p and d_p.

The derived *B/D boundary* is based on the notion of employing ε_{mb} as a general measure of relative instability. This concept has direct physical significance for Group A powders, where increasing values of ε_{mb} above ε_{mf} indicate progressively longer regions of homogeneous fluidization before the bubbling point is reached, with correspondingly progressive improvements in the fluidization quality in the bubbling regime (Kwauk, 1992). For Group B powders, however, the *particle bed* model leads to predictions of ε_{mb} values lower than ε_{mf}, which are therefore unrealizable in practice as they would imply void fractions lower than fixed bed levels. As we have seen in Chapter 8, this outcome merely translates into the physical situation of a powder that bubbles right from the onset of fluidization, so that in practice it is customary to write: $\varepsilon_{mb} = \varepsilon_{mf}$. However, it seems reasonable to assume that a system *predicted* to become unstable at a void fraction much lower than ε_{mf} could be expected to manifest greater evidence of instability from the onset of fluidization than one for which the stability limit is delayed to nearer that point.

The above discussion identifies predicted values of ε_{mb} as a measure of relative instability, and hence of fluidization quality, in bubbling systems. In the following section the significance of this relation to perturbation propagation in fluidized beds will be examined. For the moment it is sufficient to point out that predicted ε_{mb} values lower than about 0.1 correspond approximately to the Group D powders of the Geldart map: the predicted B/D boundary, shown in Figure 10.1, represents the locus of (d_p, ρ_p) combinations which result in computed values of ε_{mb} of exactly 0.1.

Although we have dealt here solely with predictions for ambient air fluidization, for which validation by means of the counterpart empirical relations may be readily confirmed, the procedures outlined are quite generally applicable. The immediate conclusion is that predicted ε_{mb} values provide a continuous measure of fluidization quality across the whole spectrum of behaviour corresponding to the Geldart powder classification map. However, when it comes to the general situation of fluidization *by any fluid,* this measure turns out to be by no means complete. To appreciate this point it becomes necessary to examine in more detail the perturbation wave relations that delivered the ε_{mb} predictions in the first place. This will then lead to more comprehensive predictive criteria for fluidization quality in general.

Perturbation propagation in fluidized beds

We now examine the propagation of void fraction perturbation-waves through fluidized beds by means of the solution to the linearized *particle bed* model equations, which gave rise to the stability criterion considered in some detail in Chapters 6, 8 and 9. In the idealized situation of perturbation-free fluidization, all systems could fluidize homogeneously, the particles motionless, the fluid pressure gradients unchanging in all points of the bed. In practice, the continual generation and propagation of void fraction disturbances give rise to continual particle motion and fluid pressure fluctuations in even the most stable configurations. The condition of fluid-dynamic stability does not imply motionless particles, but rather that their chaotic behaviour is contained, inhomogeneities decaying with time at rates which vary from system to system. This decay-rate variation could be in part responsible for differences in fluidization quality observed in *homogeneously* fluidized beds, although other factors come into play with these systems, which we will consider in the final section of this chapter.

For fluid-dynamically *unstable* systems, the behaviour spectrum is more extreme: we shall see that *large* perturbation-amplitude growth-rates can lead to the virtually instantaneous formation of complete voids (bubbles) low in the bed, where distributor-induced void fraction disturbances are inevitable occurrences. These voids rise rapidly, growing and coalescing as they go, carrying particles with them, and bursting through the bed surface to create the vigorously boiling-liquid appearance typical of gas-fluidization. At the other extreme of unstable behaviour, *low* perturbation-amplitude growth rates will be seen to be associated with the horizontal bands of suspension, of marginally higher void fraction than the bed average, which have been observed rising slowly, virtually intact, through certain liquid-fluidized suspensions: these are discussed in Chapters 9 and 12.

The above speculations, linking fluidization quality to perturbation-amplitude growth and decay rates, are now examined. Although bubble-related phenomena clearly imply conditions outside the linear response limit of the system, *initial* growth rates, obtainable from the linearized relations, can be so large in these cases that they could be expected to play a major role in subsequent developments. The linearized particle bed model delivers explicit relations for perturbation-wave velocity and amplitude growth rate, thereby enabling the above considerations to be placed on a quantitative footing. We consider first the wave velocity v, relating it to wavelength λ, void fraction ε, and system stability; then a second general criterion for fluidization quality is derived in terms of the amplitude growth rate characteristics at the critical, minimum-bubbling condition.

Perturbation-wave velocity

The velocity v of a void fraction perturbation wave, eqn (7.17), in a fluidized bed is related to wave number k by eqn (8.33). Writing this explicitly for the positive, real root of v, and in terms of wavelength λ ($\lambda = 2\pi/k$) we have:

$$v = \sqrt{\frac{-(V_D^2 - u_D^2)}{2} + \frac{\left((V_D^2 - u_D^2)^2 + 4V_D^2 u_K^2\right)^{0.5}}{2}}, \tag{10.1}$$

where

$$V_D = \frac{3u_D^2}{8\pi u_K \varepsilon_0} \cdot (\lambda/d_p). \tag{10.2}$$

From eqn (10.1) it may be readily verified that for short and long wavelengths, $\lambda \to 0$ and $\lambda \to \infty$, the wave velocity v approaches that of the dynamic wave u_D and kinematic wave u_K respectively. The stability analysis reported in Chapter 8 showed that these limits represent velocity bounds for all wavelengths. For intermediate values of λ the wave velocity changes monotonically between these limits, either increasing or decreasing with increasing λ according to the system stability.

Figure 10.2 illustrates this behaviour. It shows perturbation-wave velocities, as functions of scaled wavelength λ/d_p, for ambient air fluidization of 70 μm alumina particles. This is a system that switches from the stable to the unstable state at a void fraction of approximately 0.52. The figure on the left represents a stable condition, $u_K < u_D$, at $\varepsilon = 0.44$; that on the right an unstable condition, $u_K > u_D$ at $\varepsilon = 0.64$. The region over which the perturbation-wave velocity differs appreciably from the limiting values of u_D or u_K is from λ/d_p values of about 1 to 100.

Figure 10.3 shows how, for the same system, perturbation-wave velocities change as the bed is progressively expanded from the point of minimum fluidization ($\varepsilon_{mf} \approx 0.4$). It shows the convergence, for waves of

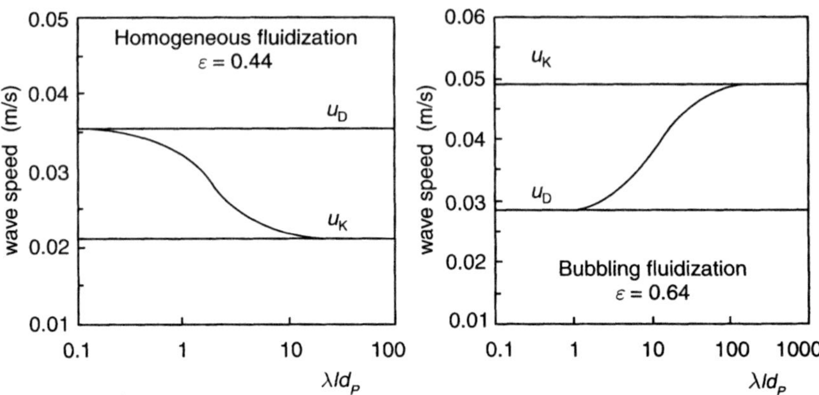

Figure 10.2 Perturbation-wave velocities as functions of wavelength: air fluidization of 70 µm alumina. ($\varepsilon_{mb} = 0.52$, $\rho_f = 1.2\,\text{kg/m}^3$, $\mu_f = 1.8 \times 10^{-5}\,\text{Ns/m}^2$, $\rho_p = 1100\,\text{kg/m}^3$, $d_p = 70\,\mu\text{m}$).

all wavelengths, to the same velocity at the minimum bubbling point: $v = u_D = u_K$, at $\varepsilon = \varepsilon_{mb}$. This general property of the minimum bubbling condition serves to illuminate its pivotal role in fluidization dynamics, a position utilized both in the previous and the following sections for the characterization of fluidization quality.

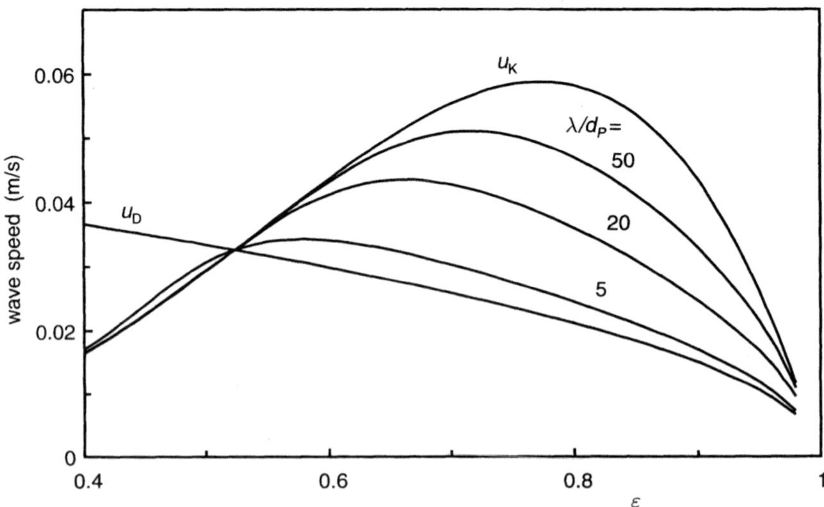

Figure 10.3 Perturbation-wave velocities as functions of wavelength: air fluidization of 70 µm alumina.

Amplitude growth rate

The amplitude growth rate a of a perturbation wave, eqn (8.32), is related to the wave velocity v, eqn (10.1), by:

$$a = \frac{3u_D^2}{4d_p u_K \varepsilon_0 v}(u_K - v). \tag{10.3}$$

For the limiting case of short wavelengths, $\lambda \to 0$, we have seen that the wave velocity approaches that of the dynamic wave, $v \to u_D$, so that eqn (10.3) becomes:

$$a|_{\lambda \to 0} = \frac{3u_D}{4d_p u_K \varepsilon_0}(u_K - u_D). \tag{10.4}$$

Figure 10.4 shows illustrative examples of this relation for the four powder groups – *A*, *B*, *C* and *D* – of the Geldart classification for ambient air fluidization, Figure 10.1.

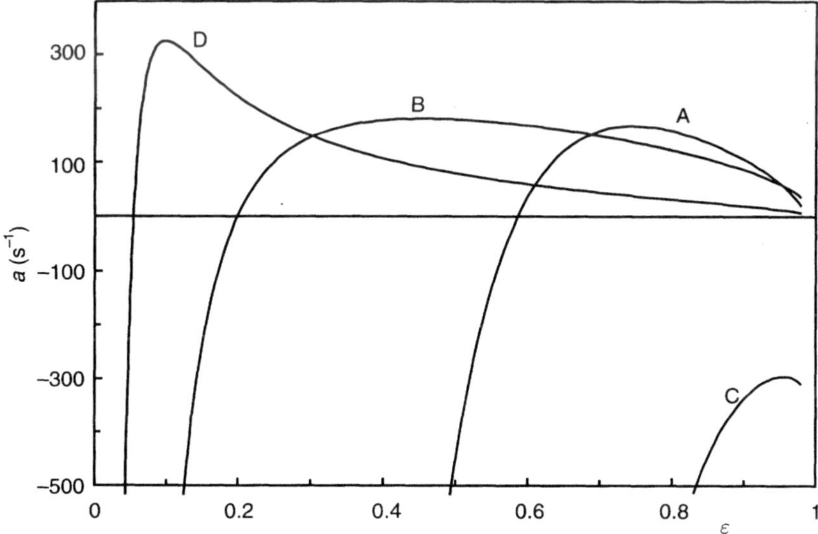

Figure 10.4 Amplitude growth rates for short wavelengths: $\lambda \to 0$, ambient air fluidization; illustrative examples for the Geldart powder classification groups:
A: $\rho_p = 1000\,\text{kg/m}^3$, $d_p = 60\,\mu\text{m}$
B: $\rho_p = 2000\,\text{kg/m}^3$, $d_p = 200\,\mu\text{m}$
C: $\rho_p = 1000\,\text{kg/m}^3$, $d_p = 20\,\mu\text{m}$
D: $\rho_p = 1000\,\text{kg/m}^3$, $d_p = 5\,\text{mm}$.

The curves shown in Figure 10.4 cut the ε axis at the 'predicted' ε_{mb} values, which we have seen provide a first effective measure of fluidization quality for ambient air-fluidized systems. The group A example cuts within the physically realizable range, $1 > \varepsilon_{mb} > 0.4$, whereas for the group C case $a|_{\lambda \to 0}$ (and, indeed, growth rates for all wavelengths) never attains the zero value that corresponds to a minimum bubbling condition. The groups B and D examples exhibit unrealizable ε_{mb} values within the ranges $0.1 \to 0.4$ and $0 \to 0.1$ respectively.

Figure 10.5 shows the general relation of eqn (10.3) as a function of wavelength for the illustrative example considered previously: air fluidization of 70 μm alumina particles under stable ($\varepsilon = 0.44$) and unstable ($\varepsilon = 0.64$) conditions. These demonstrate that for wavelengths shorter than about 100 d_p, wave amplitudes decay very fast for the stable case (left-hand figure) and grow very fast for the unstable case (right-hand figure).

The abruptness of this extensive switch in stability, brought about by simply expanding the bed across the minimum bubbling point, becomes even more clear from Figure 10.6. This shows amplitude growth rate in the same system for short wavelengths, $\lambda \to 0$ (the most sensitive to changing conditions), as a function of void fraction. The steep gradient of a with ε at the minimum bubbling point ($\varepsilon_{mb} = 0.52$, $a = 0$) indicates a system that switches from a very stable to a very unstable condition in the immediate region of ε_{mb}.

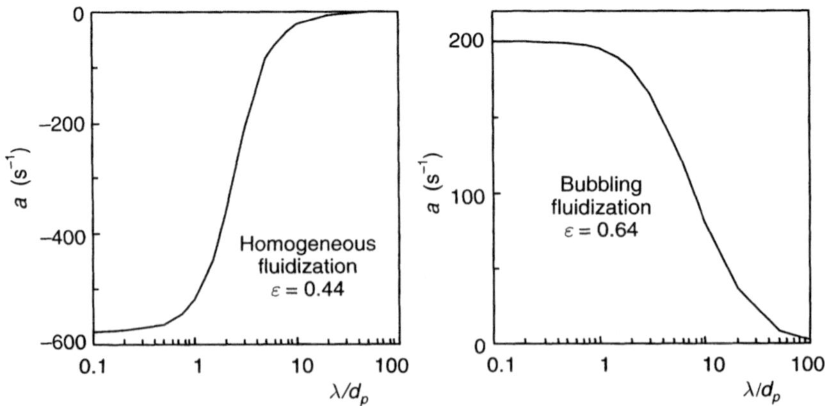

Figure 10.5 Amplitude growth rates as functions of wavelength: air fluidization of 70 μm alumina.

This last statement may be placed on a quantitative footing by first evaluating the gradient of a at ε_{mb}: $\partial a/\partial\varepsilon \approx 2000\,\mathrm{s}^{-1}$; then considering a bed expanded from just inside the stable region (at $\varepsilon = 0.519$, say) to just inside the unstable region ($\varepsilon = 0.521$), giving rise to a change in a from about -2 to $+2$. From eqn (7.17) we see that the amplitude of the perturbation wave is proportional to $\exp(at)$; so that we have here a wave which just before minimum bubbling point decays to about 1/7th of its amplitude in one second, and just beyond the minimum bubbling point increases in amplitude in one second by a factor of about 7. This indicates a system with a sharply defined minimum bubbling point, as is observed in practice.

Also represented in Figure 10.6 are two ambient water-fluidized beds (of lead-glass and copper particles) for which the particle diameters have been chosen to result in approximately the same value of ε_{mb} for all three cases. The gradient at ε_{mb}, $\partial a/\partial\varepsilon$, for the water–glass system is small this time, representing changes in amplitude (a decay just before, and a growth just after, ε_{mb}) of approximately 5 per cent per second. This corresponds to observations of water-fluidized glass particle beds, which

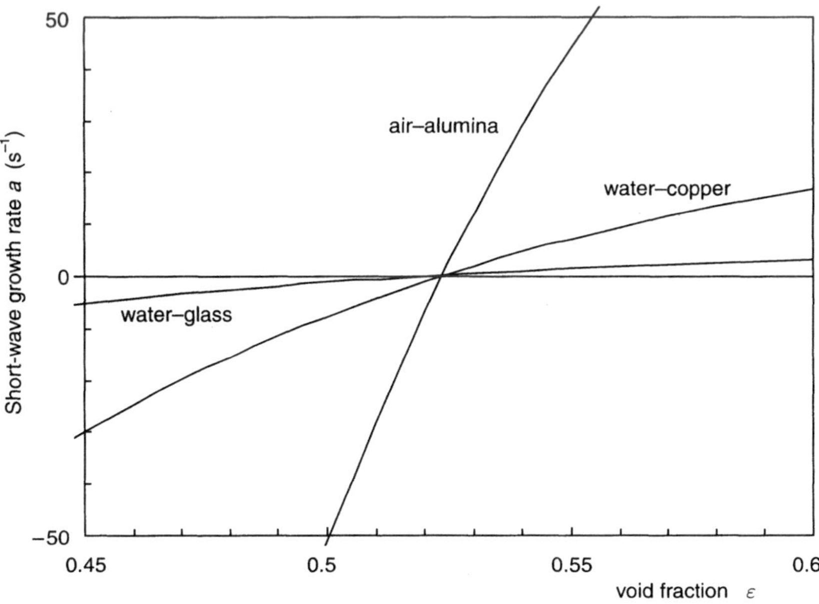

Figure 10.6 Short-wave amplitude growth rates for three systems all having $\varepsilon_{mb} \approx 0.52$ (air – 70 μm alumina; water – 5 mm lead-glass; water – 0.44 mm copper).

are discussed in Chapter 12: poorly defined transitions to the unstable state, and the persistence of mild, distributor-generated perturbations that change little during their passage through the bed.

The water–copper system falls well between the other two. This also corresponds to qualitative observations of fluidization quality in such systems. Although transition points are relatively sharp, the bubbles that result remain very small and there is no evidence of *metastable* behaviour, referred to in Chapter 9 with reference to the phenomenon of *premature bubbling* in gas-fluidized systems, and analysed in Chapter 14 on the basis of the unlinearized *particle bed* model equations; the manifestations of instability remain significantly less pronounced than is the case for the air–alumina system.

A further criterion for fluidization quality

The above discussion identifies the growth-rate gradient of short waves, $\partial a/\partial \varepsilon|_{\lambda \to 0}$, evaluated at ε_{mb}, as a further measure of fluidization quality. It provides the necessary additional dimension to the quantification in terms of the minimum bubbling void fraction, distinguishing between systems having the same ε_{mb} but different perturbation-amplitude growth rate characteristics. This gradient may be readily evaluated from eqn (10.4).

$$\left.\frac{\partial a}{\partial \varepsilon}\right|_{\varepsilon_{mb},\, \lambda=0} = 0.67 \cdot \left(\frac{g}{d_p}\right)^{0.5} \cdot \frac{2(n-1) - \varepsilon_{mb}(2n-1)}{\varepsilon_{mb}^2 (1-\varepsilon_{mb})^{0.5}}. \tag{10.5}$$

For strongly unstable systems, for which the model delivers very small, unrealizable ε_{mb} values, the ε_{mb}^2 term in the denominator of eqn (10.5) renders it unduly sensitive, leading to large changes in $\partial a/\partial \varepsilon$ with small variations in ε_{mb} which do not reflect correspondingly large variations in fluidization quality. This sensitivity problem can be significantly reduced by defining the further fluidization quality parameter Δa as the product of the gradient expression of eqn (10.5) and ε_{mb}:

$$\Delta a = \varepsilon_{mb} \left.\frac{\partial a}{\partial \varepsilon}\right|_{\varepsilon_{mb},\, \lambda=0}, \tag{10.6}$$

leading to

$$\Delta a = 0.67 \cdot \left(\frac{g}{d_p}\right)^{0.5} \cdot \frac{2(n-1) - \varepsilon_{mb}(2n-1)}{\varepsilon_{mb} (1-\varepsilon_{mb})^{0.5}}. \tag{10.7}$$

As is the case for ε_{mb}, the parameter Δa may be readily evaluated for any defined system. Because it reflects the amplitude growth-rate gradient of perturbations at the critical void fraction that separates stable from unstable fluidization, it provides a direct measure both of the extent of instability for unstable systems and the robustness of the stability manifested by those stable systems for which a minimum bubbling point exists. Systems that fluidize homogeneously for all void fractions cannot be

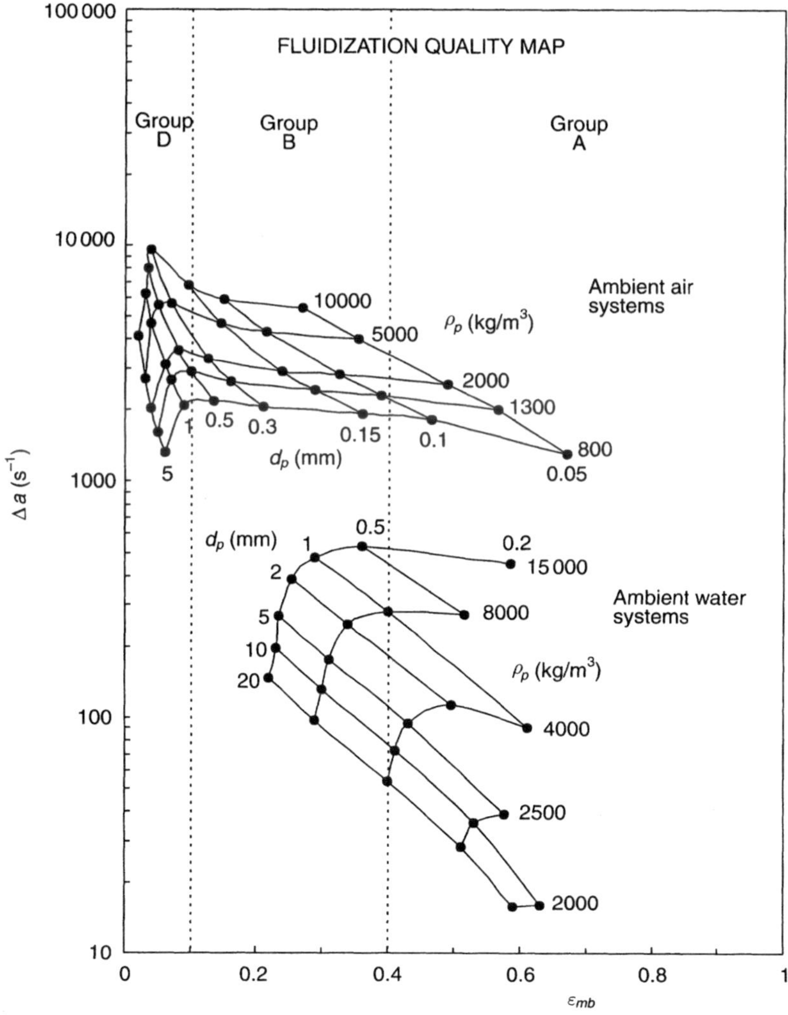

Figure 10.7 A general, predictive map of fluidization quality. Reference regions: ambient air and ambient water fluidization.

characterized by means of this parameter; these are treated separately, by means of a more appropriate fluid-dynamic criterion, in the final section of this chapter. However, it is for unstable systems that a means of predicting the quality of the ensuing fluidization is more usually required.

It has been shown above that all unstable systems may be characterized in terms of a theoretical ε_{mb}, even if values below about 0.4 have no direct physical significance. We have now an additional characterizing parameter Δa, which differentiates between systems of differing fluidization quality having the same ε_{mb}. Both relate directly to the critical, minimum-bubbling condition separating stable from unstable fluidization. Together they may be used to construct a general predictive map of fluidization quality, applicable to any defined fluidized system. This is shown in Figure 10.7 for the much-studied case of ambient air fluidization of an extensive range of particle species; also included in the map are cases of ambient water fluidization of relatively high-density particles, for which the single-phase approximation, based on the assumption that $\rho_p \gg \rho_f$, may still be tentatively applied. These examples provide reference regions that enable initial assessments to be made of previously untested fluidized systems.

Example applications

The map shown in Figure 10.7 contains examples of systems whose fluidization quality is well known. It represents a starting point that can be progressively augmented by experimental study to fill the uncharted regions. Two examples of applications illustrate how such a map may be used to obtain rapid estimates of fluidization quality in hitherto untested systems.

Example 1

Consider first the high-pressure gas fluidization ($\rho_f = 110\,\mathrm{kg/m^3}$, $\mu_f = 2 \times 10^{-5}\,\mathrm{Ns/m^2}$ – corresponding to a pressure of over 100 bar) of 200 μm particles of density $800\,\mathrm{kg/m^3}$. For this system, the stability criterion (Table 8.1) and eqn (10.7) yield values: $\varepsilon_{mb} = 0.5$, $\Delta a = 430\,\mathrm{s^{-1}}$. On the fluidization quality map, Figure 10.7, this corresponds closely to *ambient water fluidization* of 300 μm particles of density $13\,000\,\mathrm{kg/m^3}$. This finding, on the effect of high gas pressure on the fluidization of relatively low-density particles, is in line with the experimental observations of Jacob and Weimer reported in the previous chapter (Jacob and Weimer, 1987):

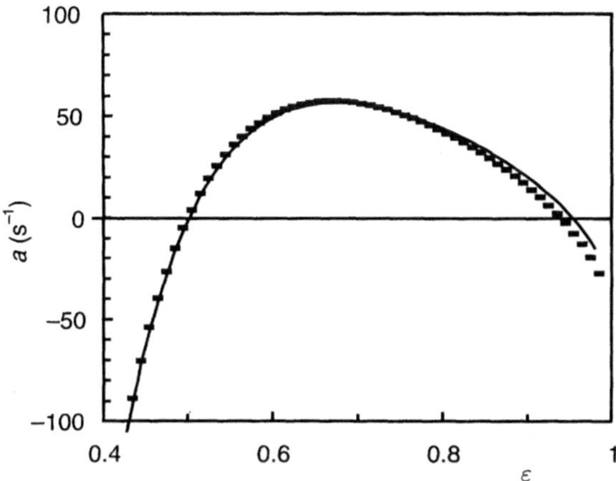

Figure 10.8 High-frequency perturbation-amplitude growth rate a as a function of void fraction ε: high pressure gas fluidization, points; 'equivalent' ambient water fluidization, continuous curve.

their systems behaved very much like water-fluidized high-density powders, which, unlike usual gas-fluidized systems, contract little (if at all) on attaining the minimum bubbling condition: this behaviour is discussed and analysed in Chapter 14.

As a further confirmation of this correspondence, eqn (10.4) may be used to plot the high-frequency, perturbation-amplitude growth rate as a function of void fraction for both systems. This is shown in Figure 10.8: the two relations remain virtually identical over the full working range.

Example 2

Consider this time the fluidization of 100 μm particles of density 2000 kg/m^3 by a gas, at high temperature and moderately elevated pressure, having a viscosity of 4×10^{-5} Ns/m^2 and a density of 1.3 kg/m^3. This leads to: $\varepsilon_{mb} = 0.47$, $\Delta a = 1930\,\text{s}^{-1}$. Referring to the fluidization quality map, it will be seen that these values correspond closely to *ambient air-fluidization* of 100 μm particles of density 800 kg/m^3 – a typical alumina catalyst. Once again, the two parameters, ε_{mb} and Δa, appear to provide excellent characterization of amplitude growth rates over the full expansion range: the counterpart of the Figure 10.8 comparison for the previous example also showing these two systems to be being virtually identical in this respect.

Homogeneous fluidization

It is well known that, whereas the particles in homogeneous liquid-fluidized beds usually exhibit considerable random motion, this is not the case for gas-fluidized particles, which are much more firmly held together in suspension. This difference in behaviour has been the subject of long-running imaginative speculation, generally invoking the presence of extraneous, non-fluid-dynamic interactions between gas-fluidized particles, which are supposedly absent in liquid systems (Martin, 1983).

An early experimental study (Rietema and Mutsers, 1973) appeared to support this hypothesis: it reported that the surface of a homogeneously fluidized gas bed that had been tilted somewhat from the vertical orientation itself displayed a tilt – as though the particles had been glued together by contact forces into a rigid structure. However, a more detailed study (Gilbertson and Yates, 1997) showed this effect to be due to the non-vertical flow of gas, resulting in '... a mechanical structure within the bed after partial defluidization rather than the action of interparticle forces'.

A more recent experimental study (Marzocchella and Salatino, 2000) into the effect of pressure on bed stability repeats the particle–particle contact bond hypothesis for homogeneous gas fluidization on the sole basis of observations of lower particle random motion than is the case for liquid fluidization; no clue is provided as to the nature of the postulated interparticle forces, nor for why they should strengthen with increasing pressure – the major conclusion of the investigation. In fact, the results presented are in very reasonable quantitative accord with the fluid-dynamic stability criterion of eqn (8.36), which predicts directly the observed increase in ε_{mb} with fluid pressure. The fact that the interparticle force hypothesis leads to no quantitative (or even, in cases such as the one just quoted, qualitative) predictions of bed stability perhaps helps to explain its longevity; this feature makes it very difficult to disprove.

It is clear from previous sections of this chapter that fluid-dynamic phenomena can account for a wide spectrum of behaviour patterns in unstable heterogeneous fluidized beds, and there is no reason to suppose that the same should not be so for stable homogeneous systems. The problem becomes that of identifying a relevant criterion for characterizing such differences. As no minimum-bubbling condition is predicted for fully homogeneous beds, fluidization quality criteria based on ε_{mb} and Δa are inapplicable in these cases.

The *bulk mobility* B_p of the particles

Consider a homogeneously fluidized bed in equilibrium. If the particles are now subjected to a small force, they will move to restore the equilibrium condition. How fast they do this will depend on the specific system properties: the greater the velocity of the particles, the more uniformly held together will be the suspension, and *vice versa*. A parameter that could provide a measure of this effect, the '*bulk mobility* B_p of the particles', has been proposed by Batchelor (1988) in the development of a model for fluidization that is structurally similar to the particle bed model. He defines B_p as: 'the ratio of the (small additional) mean velocity, relative to zero-volume-flux axes, to the (small additional) steady force applied to each particle of a homogeneous dispersion'. For a bed initially in a state of equilibrium, this becomes:

$$B_p = \left.\frac{\partial u_p}{\partial f}\right|_{f=0}. \tag{10.8}$$

B_p may be readily evaluated from eqn (8.1), the expression for the net primary force acting on a fluidized particle in equilibrium. From this we obtain:

$$\frac{\partial f}{\partial u_p} = -\frac{0.8\pi d_p^3 g(\rho_p - \rho_f)\varepsilon^{-3.8}}{n u_t} \cdot \left(\frac{U_0 - u_p}{u_t}\right)^{\frac{4.8}{n}-1}; \tag{10.9}$$

leading to, for the equilibrium condition ($f = 0$, $u_p = 0$, $U_0 = u_t\varepsilon^n$):

$$B_p = \left.\frac{\partial u_p}{\partial f}\right|_{f=0} = \frac{1.25 n u_t \varepsilon^{n-1}}{\pi d_p^3(\rho_p - \rho_f)g}. \tag{10.10}$$

The mobility number Mo

Equation (10.10) may be expressed in dimensionless form, thereby defining the *mobility number Mo*:

$$Mo = \left(B_p \mu_f d_p\right) = \frac{0.40 n \varepsilon^{n-1} Re_t}{Ar}. \tag{10.11}$$

As both Re_t and n are functions of the Archimedes number Ar, eqns (2.17) and (4.5) respectively, Mo may be expressed as a function of Ar and void fraction ε. This is shown in Figure 10.9 over the relevant range for homogeneous fluidization.

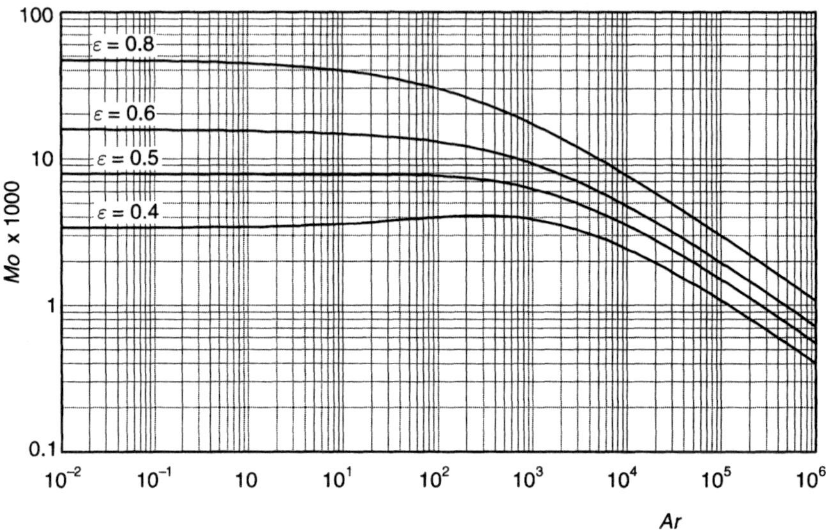

Figure 10.9 The *mobility number Mo* for homogeneously fluidized particles, as a function of *Archimedes number Ar*.

Note that under low Reynolds number conditions we have that $n = 4.8$, and, from eqn (2.15), that $Ar/Re_t = 18$, so that Mo becomes a function solely of void fraction:

$$Mo = 0.107\varepsilon^{3.8}, \qquad B_p = \frac{0.107\varepsilon^{3.8}}{\mu_f d_p}. \tag{10.12}$$

These relations are reflected in the initial, horizontal sections of the curves in Figure 10.9. They apply to most gas-fluidized beds, which operate in the low Reynolds number regime. Equation (10.12) indicates at a glance why it is that low-viscosity, small particle size, gas-fluidized beds exhibit high particle bulk mobilities, and, as a consequence, are more firmly held together under homogeneous fluidization conditions than are liquid beds.

To make a specific comparison, consider the following typical homogeneous beds: air-fluidized 80 μm alumina and water-fluidized 1 mm glass, both at void fractions of 0.5. Equation (10.10) delivers B_p values of 5 570 000 m/Ns and 3170 m/Ns respectively; the bulk mobility of the gas-fluidized particles is nearly 2000 times greater than that of the liquid-fluidized ones. It is unsurprising, therefore, that the comportment of the two beds should be so different.

References

Batchelor, G.K. (1988). A new theory for the instability of a uniform fluidized bed. *J. Fluid Mech.*, **183**, 75.

De Luca, L., Hellenbroich, D., Titchener-Hooker, N.J. and Chase, H.A.A. (1994). A study of the expansion characteristics and transient behaviour of adsorbent particles suitable for bioseparations. *Bioseparation*, **4**, 311.

Geldart, D. (1973). Types of fluidization. *Powder Technol.*, **7**, 275.

Gibilaro, L.G., Di Felice, R. and Foscolo, P.U. (1988). On the minimum bubbling and the Geldart classification for gas-fluidized beds. *Powder Technol.*, **56**, 21.

Gilbertson, M.A. and Yates, J.G. (1997). Bubbles, jets, X-rays and nozzles: what happens to fluidized beds under pressure. *Inst. Chem. Eng. Jubilee Research Event*, 433.

Jacob, K.V. and Weimer, A.W. (1987). High-pressure particulate expansion and minimum bubbling of fine carbon powders. *AIChE J.*, **33**, 1698.

Kwauk, M. (1992). *Fluidization: Idealized and Bubbleless, with Applications*. Ellis Horwood.

Martin, P.D. (1983). On the particulate and delayed bubbling regimes in fluidization. *Chem. Eng. Res. Des.*, **61**, 318.

Marzocchella, A. and Salatino, P. (2000). Fluidization of solids with CO_2 at pressures from ambient to supercritical. *AIChE J.*, **46**, 901.

Rietema, K. and Mutsers, S.M.P. (1973). The effect of interparticle forces on the expansion of a homogeneous gas-fluidized bed. *Proc. Int. Symp. Fluidization and its Applications*. Toulouse, France.

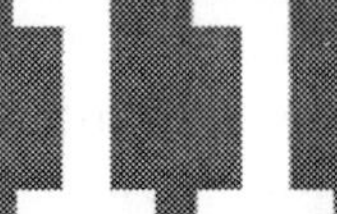

The two-phase particle bed model

The fluid pressure field

The analysis presented in Chapter 8 was solely in terms of the conservation equations for the *particle phase* of a fluidized suspension. However, the full one-dimensional description is in terms of the coupled mass and momentum conservation equations for both the *particle* and *fluid* phases: eqns (8.21)–(8.24). These equations correspond to those derived in Chapter 7, except for the inclusion of the particle-phase elasticity term on the extreme right of eqn (8.22).

The decoupling of these separate phase descriptions, which enabled the particle phase to be treated independently, involved two approximations of the phase interaction term. The first one was set out in Chapter 7: both particles and fluid were regarded as being incompressible. This was justified on the basis that only a gas phase is going to exhibit any significant compressibility, and the orders-of-magnitude differences in particle and fluid densities for gas fluidization render quite insignificant the small changes in gas density resulting from compression. This assumption led to the relation linking fluid- and particle-phase velocities at all locations:

$$U_0 = \varepsilon u_f + (1 - \varepsilon)u_p, \tag{11.1}$$

where U_0 is the entering fluid flux. Equation (11.1) enables the fluid velocity variable u_f to be expressed in terms of the particle velocity u_p at all points in the bed.

The second approximation was to regard the particle density as being much greater than the fluid density: $\rho_p \gg \rho_f$. This condition is certainly applicable to almost all gas-fluidized beds, but not many liquid-fluidized ones. It enabled the fluid pressure gradient, $\partial p/\partial z$, to be approximated in terms of particle drag, eqn (8.25). Fluid pressure p is a further variable (along with u_p, u_f and ε) in the describing equations, the fluid pressure gradient appearing directly as a surface force in the fluid-phase momentum equation, eqn (8.24); it also determines particle buoyancy, eqn (4.17). We now do away with this second approximation, making no assumption whatsoever concerning the fluid pressure field.

The combined momentum equation

The operations we now describe constitute a specific case (Foscolo *et al.*, 1989) of the general procedure proposed by Wallis (1969) for one-dimensional, two-phase systems. It consists of combining the fluid and particle momentum equations, eqns (8.22) and (8.24), by elimination of the fluid pressure gradient, which appears in both of them. This yields the combined momentum equation:

$$\rho_p\left(\frac{\partial u_p}{\partial t} + u_p\frac{\partial u_p}{\partial z}\right) - \rho_f\left(\frac{\partial u_f}{\partial t} + u_f\frac{\partial u_f}{\partial z}\right) = \frac{F_d}{\varepsilon(1-\varepsilon)} - g(\rho_p - \rho_f) + \frac{\rho_p u_{\mathrm{D}}^2}{(1-\varepsilon)}\cdot\frac{\partial \varepsilon}{\partial z}. \tag{11.2}$$

Equation (11.2), together with the continuity equations for the two phases, eqns (8.21) and (8.23), now define the two-phase system – taking full account of fluid pressure variation.

Stability analysis

We now proceed exactly as in Chapters 7 and 8. The system equations are first linearized about the steady-state, equilibrium condition: $\varepsilon = \varepsilon_0$,

$u_f = u_{f0}$, $u_p = u_{p0} = 0$. Then the perturbation-wave solution is examined to see whether or not small variations in void fraction ε^* will start to grow or decay.

The linearized equations of change

Particle-phase continuity, eqn (8.21), and the combined momentum equation, eqn (11.2), become respectively on linearization:

$$\frac{\partial u_p}{\partial z} = \frac{1}{(1-\varepsilon_0)} \cdot \frac{\partial \varepsilon^*}{\partial t}, \tag{11.3}$$

$$\frac{\partial u_p}{\partial t} = \frac{1}{\varepsilon_0 \rho_p + (1-\varepsilon_0)\rho_f} \cdot \left[-\rho_f u_{f0} \frac{\partial \varepsilon^*}{\partial t} - \left(\rho_f u_{f0}^2 - \frac{\rho_p u_D^2 \varepsilon_0}{(1-\varepsilon_0)} \right) \frac{\partial \varepsilon^*}{\partial z} \right.$$
$$\left. - \frac{4.8}{n u_{f0}} \cdot g(\rho_p - \rho_f) u_p \; - 4.8 g(\rho_p - \rho_f)\varepsilon^* \right]. \tag{11.4}$$

Linearization of combined momentum equation

Expressing eqn (11.2) in terms of deviation variables (ε^*, u_p, u_f^*), expanding about the steady-state condition (ε_0, 0, u_{f0}), and retaining only linear terms yields:

$$\rho_p \cdot \frac{\partial u_p}{\partial t} - \rho_f \left[\frac{\partial u_f^*}{\partial t} + u_{f0} \frac{\partial u_f^*}{\partial z} \right] =$$
$$\left(\frac{F_{d0} - F_{d0}(1-\varepsilon_0)\varepsilon^* + \mathrm{f}_\varepsilon \varepsilon^* + \mathrm{f}_{u_p} u_p}{\varepsilon_0(1-\varepsilon_0)} \right)$$
$$- g(\rho_p - \rho_f) + \frac{\rho_p u_D^2}{(1-\varepsilon_0)} \cdot \frac{\partial \varepsilon^*}{\partial z}.$$

Eqn (11.4) emerges on inserting the following relations into the above equation: the steady-state condition,

$$F_{d0} = \varepsilon_0(1-\varepsilon_0) g(\rho_p - \rho_f);$$

relations for the partial derivatives,

$$\mathrm{f}_\varepsilon = \partial F_d / \partial \varepsilon |_0 = -3.8(1-\varepsilon_0)(\rho_p - \rho_f) g,$$
$$\mathrm{f}_{u_p} = \partial F_d / \partial u_p |_0 = -4.8(1-\varepsilon_0)(\rho_p - \rho_f) g / n u_{f0};$$

$$\frac{\partial u_f^*}{\partial z} = -\frac{1}{\varepsilon_0} \cdot \frac{\partial \varepsilon^*}{\partial t} - \frac{u_{f0}}{\varepsilon_0} \cdot \frac{\partial \varepsilon^*}{\partial z}$$

(from linearized fluid continuity, eqn (8.23)),

$$\frac{\partial u_f^*}{\partial t} = -\frac{(1-\varepsilon_0)}{\varepsilon_0} \cdot \frac{\partial u_p}{\partial t} - \frac{u_{f0}}{\varepsilon_0} \cdot \frac{\partial \varepsilon^*}{\partial t}$$

(from the linearized 'incompressible phases' assumption, eqn (11.1)).

On differentiating eqn (11.3) with respect to t, and eqn (11.4) with respect to z, and then equating the resulting right-hand sides, we obtain the partial differential equation describing small void fraction perturbations in the bed:

$$\frac{\partial^2 \varepsilon^*}{\partial t^2} + 2V \frac{\partial^2 \varepsilon^*}{\partial t \partial z} + G \frac{\partial^2 \varepsilon^*}{\partial z^2} + D\left(\frac{\partial \varepsilon^*}{\partial t} + u_{\mathrm{K}} \frac{\partial \varepsilon^*}{\partial z}\right) = 0, \tag{11.5}$$

where:

$$D = \frac{4.8(\rho_p - \rho_f)g(1-\varepsilon_0)}{u_{\mathrm{K}}(\varepsilon_0 \rho_p + (1-\varepsilon_0)\rho_f)}, \tag{11.6}$$

$$u_{\mathrm{K}} = u_t n(1-\varepsilon_0)\varepsilon_0^{n-1}, \tag{11.7}$$

$$G = \frac{(1-\varepsilon_0)\rho_f u_{f0}^2 - \rho_p u_{\mathrm{D}}^2 \varepsilon_0}{\varepsilon_0 \rho_p + (1-\varepsilon_0)\rho_f}, \tag{11.8}$$

$$V = \frac{0.5\rho_f u_{f0}(1-\varepsilon_0)}{\varepsilon_0 \rho_p + (1-\varepsilon_0)\rho_f}. \tag{11.9}$$

The denominator that appears in eqns (11.8) and (11.9) looks strange; it represents a weighted density obtained by summing the particle density multiplied by the fluid fraction and the fluid density multiplied by the particle fraction. The *physical* significance of this topsy-turvy combination is unclear. The same can be said for the velocity V itself, eqn (11.9), which corresponds to the 'weighted mean velocity' defined by Wallis (1969), in which fluid velocity is weighted with particle fraction and *vice versa*. (Only the fluid velocity component appears in eqn (11.9) because for the case under consideration we have that $u_{p0} = 0$.) *Mathematically*, however, velocity V has an important significance, which will soon become clear.

Note that for gas fluidization, $\rho_p \gg \rho_f$, eqns (11.6), (11.8) and (11.9) become:

$$D = \frac{4.8g}{u_K} \cdot \frac{1-\varepsilon_0}{\varepsilon_0}, \quad G = -u_D^2, \quad V = 0, \tag{11.10}$$

thereby reducing eqn (11.5) to its single phase counterpart, eqn (8.31).

The travelling wave solution

We now proceed exactly as in Chapters 7 and 8: the partial derivatives in eqn (11.5) are evaluated from the travelling void fraction perturbation-wave expression,

$$\varepsilon^* = \varepsilon_A \cdot \exp((a - ikv)t).\exp(ikz), \tag{11.11}$$

to yield the complex algebraic equation:

$$\begin{aligned} &a^2 - k^2v^2 + 2Vk^2v - Gk^2 + Da \\ &\quad + i(-2akv + 2akV - Dkv + Dku_K) = 0. \end{aligned} \tag{11.12}$$

On equating the real and imaginary parts of eqn (11.12) to zero, we obtain, after some manipulation, expressions for the amplitude growth rate a and the square of the wave number k:

$$a = \frac{D}{2(v-V)} \cdot ((u_K - V) - (v - V)), \tag{11.13}$$

$$k^2 = \frac{D^2}{4(v-V)^2} \cdot \frac{(u_K - V)^2 - (v-V)^2}{(v-V)^2 - (V^2 - G)}. \tag{11.14}$$

The significance of the 'weighted mean velocity' V is now apparent from the forms of eqns (11.13) and (11.14): it is relative to V that system wave velocities are most naturally expressed. The only exception concerns the $(V^2 - G)$ term in the denominator of eqn (11.14). However, this too may be brought into line by *defining* a wave velocity u_{DT} such that:

$$(u_{DT} - V)^2 = (V^2 - G). \tag{11.15}$$

Making this substitution in eqn (11.14), and writing all wave velocities relative to V,

$$\hat{v} = v - V, \quad \hat{u}_K = u_K - V, \quad \hat{u}_{DT} = u_{DT} - V, \tag{11.16}$$

yields:

$$a = \frac{D}{2\hat{v}} \cdot (\hat{u}_K - \hat{v}), \tag{11.17}$$

$$k^2 = \frac{D^2}{4\hat{v}^2} \cdot \frac{\hat{u}_K^2 - \hat{v}^2}{\hat{v}^2 - \hat{u}_{DT}^2}. \tag{11.18}$$

Equations (11.17) and (11.18) are now identical in form to those obtained in the simplified, 'single-phase' treatment of Chapter 8. Equation (11.18) shows the defined quantity u_{DT} to be the velocity of short perturbation waves ($k \to \infty$, $v \to u_{DT}$) and hence the dynamic-wave velocity in the two-phase treatment.

The two-phase stability criterion

The stability condition therefore remains that uncovered in Chapter 8. To recap: eqn (11.18) indicates that for k to be real (k^2 +ve), $\hat{v}$ must lie between $\hat{u}_K$ and $\hat{u}_{DT}$; eqn (11.17) then indicates that stability (a –ve) corresponds to $\hat{u}_{DT} > \hat{u}_K$ and *vice versa*: it depends solely on the difference in relative velocities, $\hat{u}_{DT} - \hat{u}_K$, which is clearly the same as the difference in velocities relative to the stationary, steady-state particle phase, $u_{DT} - u_K$:

$$u_{DT} - u_K \begin{cases} +\text{ve} & \textit{Stable: homogeneous fluidization} \\ = 0 & \textit{Stability limit: } \varepsilon = \varepsilon_{mb} \\ -\text{ve} & \textit{Unstable: bubbling fluidization} \end{cases} \tag{11.19}$$

The kinematic-wave velocity u_K maintains the same expression as in the single-phase treatment, eqn (11.7).

The two-phase dynamic-wave velocity u_{DT}

The new upward characteristic (dynamic-wave) velocity u_{DT} may be evaluated from eqns (11.8), (11.9) and (11.15):

$$u_{DT} = \sqrt{V^2 - G} + V = \sqrt{u_D^2 P - P(1-P)u_{f0}^2} + (1-P)u_{f0}, \tag{11.20}$$

where

$$P = \frac{\rho_p \varepsilon_0}{\rho_p \varepsilon_0 + \rho_f (1 - \varepsilon_0)}, \tag{11.21}$$

and the equilibrium fluid velocity u_{f0} may be expressed in terms of the equilibrium void fraction ε_0:

$$u_{f0} = \frac{U_0}{\varepsilon} = u_t \varepsilon_0^{n-1}. \tag{11.22}$$

Note that where particle density is much larger than fluid density we have:

$$\rho_p \gg \rho_f, \quad P \to 1, \quad u_{\mathrm{DT}} \to u_{\mathrm{D}}, \tag{11.23}$$

and the stability criterion, eqn (11.19), reduces to the single phase expression, eqn (8.36), for which:

$$u_{\mathrm{D}} = \sqrt{3.2 g d_p (1 - \varepsilon_0)(\rho_p - \rho_f)/\rho_p}. \tag{11.24}$$

This result confirms once again the validity of the single-phase approximation adopted in Chapter 8 for cases of gas fluidization.

Note that eqn (11.5) may be cast directly in terms of the two-phase dynamic-wave speed. On the basis of the relation for u_{DT}, eqn (11.15), we obtain:

$$\left(\frac{\partial}{\partial t} + u_{\mathrm{DT1}}\frac{\partial}{\partial z}\right)\left(\frac{\partial}{\partial t} + u_{\mathrm{DT2}}\frac{\partial}{\partial z}\right)\varepsilon^* + D\left(\frac{\partial}{\partial t} + u_{\mathrm{K}}\frac{\partial}{\partial z}\right)\varepsilon^* = 0; \tag{11.25}$$

where

$$u_{\mathrm{DT1}} = V - \sqrt{V^2 - G}, \qquad u_{\mathrm{DT2}} = V + \sqrt{V^2 - G}. \tag{11.26}$$

Equation (11.25) represents a standard form in which u_{DT1} and u_{DT2} are the higher order characteristic (dynamic) speeds (Whitham, 1974).

References

Foscolo, P.U., Di Felice, R. and Gibilaro, L.G. (1989). The pressure field in an unsteady-state fluidized bed. *AIChE J.*, **35**, 1921.

Wallis, G.B. (1969). *One-Dimensional Two-Phase Flow*. McGraw-Hill.

Whitham, G.B. (1974). *Linear and Non-Linear Waves*. John Wiley & Sons.

12

Two-phase model predictions and experimental observations

Comparison of the one- and two-phase models

In order to place the results of Chapter 11 in a practical perspective, we first examine typical examples of air and water fluidization. Before proceeding, it may be helpful to recall that systems which exhibit a transition to bubbling behaviour at the critical void fraction ε_{mb} return to the homogeneous state at a higher void fraction (often much higher, approaching unity): there are always either two or zero solutions for the transitional void fraction, as we saw in the opening section of Chapter 9.

This is illustrated in Figure 12.1. The stability criterion is again expressed in terms of the dimensionless stability function S. Transitions between stable and unstable fluidization occur at $S = 0$:

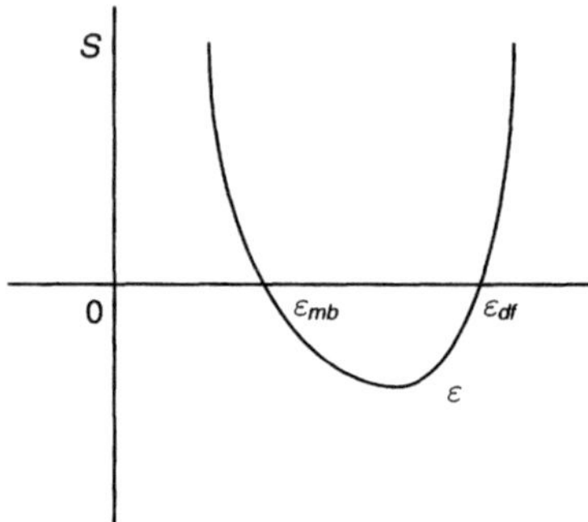

Figure 12.1 The stability function for bubbling and transitional systems.

$$S = \frac{u_{\mathrm{D}}\ (\text{or } u_{\mathrm{DT}}) - u_{\mathrm{K}}}{u_{\mathrm{K}}} = 0, \quad \text{for } \varepsilon = \varepsilon_{mb} \text{ and } \varepsilon = \varepsilon_{df}, \tag{12.1}$$

the second transition ε_{df} marks the return to homogeneous behaviour in what may be termed the 'dilute fluidization' regime.

Stability predictions for air- and water-fluidization

Table 12.1 reports illustrative values of the two transitional void fractions, ε_{mb} and ε_{df}, evaluated using the one- and two-phase dynamic-wave velocity expressions, eqns (11.24) and (11.20) respectively.

The results of Table 12.1 confirm the validity of the single-phase approximation for gas fluidization, even under very high-pressure conditions.

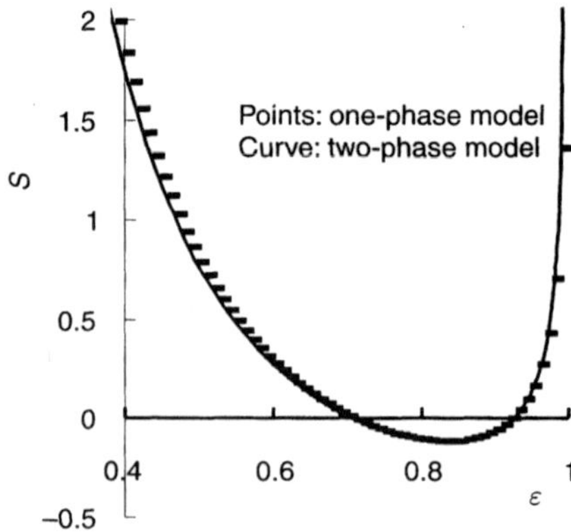

Figure 12.2 Air-fluidization: the stability function for fluidization of alumina at 50 bar ($\rho_p = 1000\,\mathrm{kg/m^3}$, $d_p = 60\,\mu\mathrm{m}$).

Table 12.1 Comparison of the one- and two-phase *particle bed* models: fluidization by air and water (at temperatures of 20 °C and ambient pressure, unless otherwise indicated)

Fluid	*Particles*			ε_{mb}		ε_{df}	
	Material	ρ_p (kg/m^3)	d_p (μm)	*One-phase model*	*Two-phase model*	*One-phase model*	*Two-phase model*
Air	Alumina	1000	100	0.45	0.45	> 0.99	> 0.99
Air (at 50 bar)	Alumina	1000	60	0.71	0.71	0.92	0.93
Air (at 50 bar)	Sand	2500	60	0.44	0.44	0.98	0.99
Water	Glass	2500	2500	0.69	0.68	0.79	0.89
Water	Copper	8700	400	0.55	0.55	0.93	0.95

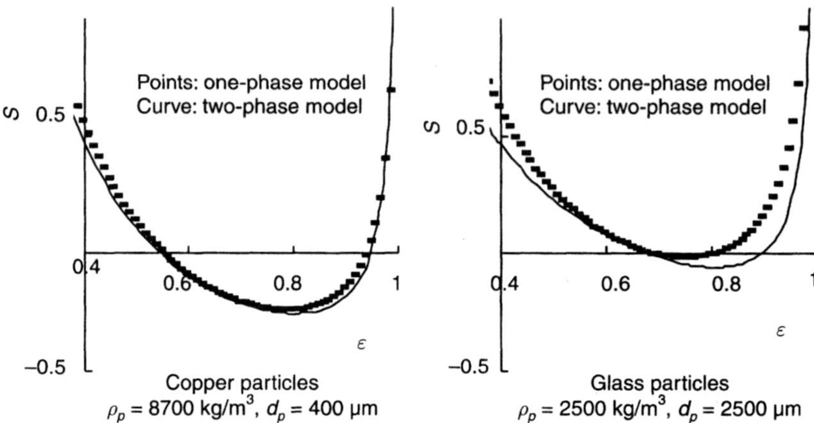

Figure 12.3 Water fluidization: the stability function for fluidization of copper and glass.

For liquid fluidization, the one-phase model still proves reasonably effective in predicting ε_{mb}, but tends to underestimate the second transitional void fraction ε_{df}, particularly for particles of relatively low density. These conclusions are further illustrated in Figures 12.2 and 12.3, which show S as a function of void fraction for three of the Table 12.1 examples.

For practical purposes, the water-fluidized copper particles may be treated by the one-phase model: the particle/fluid density ratio, although considerably less than for virtually all gas-fluidized systems, is sufficiently high for this to be the case. Water-fluidized glass, on the other hand, where the density ratio is only 2.5, exhibits a clear difference in the one- and two-phase model predictions, the former underestimating the extent of the unstable (negative S) region. Unfortunately, from the standpoint of experimental verification, this mismatch corresponds to a region of 'indeterminate stability', which is discussed below; perturbation growth/decay rates in this region are so small as to leave open the question whether or not observed behaviour corresponds to system stability or instability. It also corresponds to a region where, as the minimum in S is close to the ε axis, 'model sensitivity' problems can arise, as discussed in Chapter 9.

Dynamic-wave velocity

Differences in one-phase and two-phase model stability predictions for linearized systems are due solely to differences in the dynamic-wave

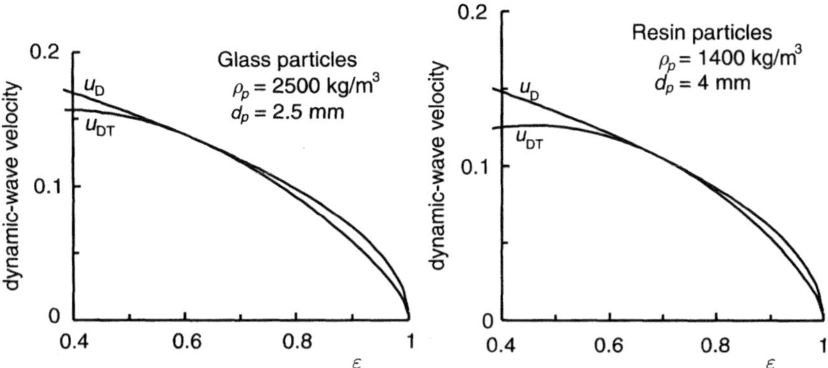

Figure 12.4 Fluidization by ambient water: comparison of one- and two-phase dynamic-wave velocities.

velocity expressions, eqns (11.20) and (11.24). For gas fluidization, these differences are negligible. For liquid systems, the examples of Figure 11.3 suggest that they could be significant for systems in which the particle density approaches that of the fluid. However, the direct comparison of one- and two-phase dynamic-wave velocities shown in Figure 12.4 would appear to indicate that, even under conditions of quite small fluid/particle density difference, the one- and two-phase models remain tolerably in agreement over most of the void fraction range, the maximum deviation occurring at the minimum fluidization condition, $\varepsilon = \varepsilon_{mf}$. The figures show the water–glass system considered above, and a water–resin system having a solid/fluid density ratio of 1.4. This latter system fluidizes homogeneously over the entire expansion range.

Liquid-fluidized systems

Predictions of the *single-phase* particle bed model were confronted with experimental observations of gas-fluidized beds in Chapter 9. The assumption of $\rho_p \gg \rho_f$, which enabled the fluid-phase equations to be effectively removed from consideration in this case, would appear to render this approximation inappropriate for most cases of liquid fluidization. The above results, however, show that the single-phase approximation leads to stability predictions in reasonable harmony with the full two-phase model for liquid fluidization over a substantial range of particle density, down to perhaps three times that of the fluid. In this section we confront reported experimental observations relating to the stability of liquid-fluidized systems with the one- and two-phase particle bed model predictions.

The stability map for fluidization by ambient water

The global stability map of Figure 12.5 was constructed exactly as described in Chapter 9, where its counterpart for fluidization by ambient air was presented. Although the two-phase relation for the dynamic-wave velocity u_{DT}, eqn (11.20), was used in this case, the results shown are virtually indistinguishable from those resulting from the single-phase formulation for u_D, eqn (11.24). It will be noticed that the relevant particle sizes are larger than for the ambient air-fluidization map, and it is the left-hand boundary, that separating 'normal' liquid beds that fluidize homogeneously from those exhibiting a transition to bubbling behaviour, which is of more practical value.

The significance of this particular stability map can be enhanced by considering it in parallel with the relevant perturbation-amplitude growth rates discussed in Chapter 10. For the higher density powders, for which perturbation growth rates are relatively high, a well-defined transition from homogeneous to bubbling behaviour is found to occur: these systems display many similarities with gas-fluidized beds. For lower density powders the situation is less clear: a well-defined transition from homogeneous behaviour is not observed, nor are completely void bubbles formed. Instead, the area close to and to the right of the homogeneous-transition region boundary is characterized by inhomogeneities that can take the form of horizontal, high void fraction bands which rise slowly

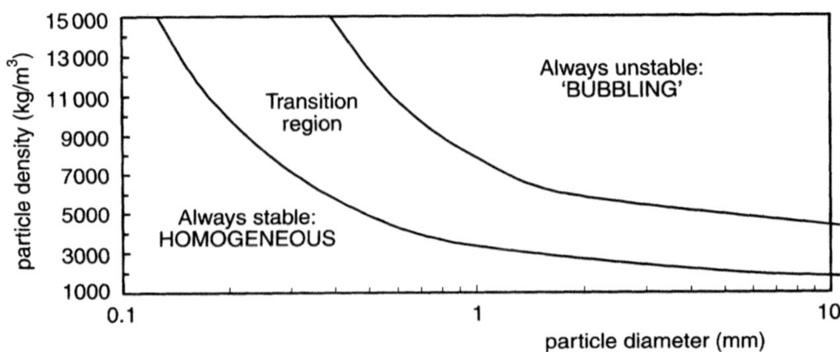

Figure 12.5 Stability map for fluidization by ambient water.

through the bed, and which may be related to predictions of perturbation growth rates that are very much smaller than those encountered in 'bubbling' systems. This phenomenon was referred to in Chapter 10 with regard to the gas- and liquid-fluidized systems featured in Figure 10.6, which manifested large differences in perturbation amplitude growth rates in the vicinity of the minimum fluidization point.

Indeterminate stability

Fluidized systems for which perturbation-amplitude decay rates or growth rates are small can display persistent, distributor-generated void fraction inhomogeneities which change little during their passage through the bed; under these circumstances it may be difficult to tell whether the observed behaviour represents stable or unstable fluidization. This phenomenon has been studied experimentally for water-fluidized beds, and related to persistently low absolute values of the growth-rate parameter a predicted by the *particle bed* model (Gibilaro *et al.*, 1988). This has enabled the stability map for ambient water-fluidization (Figure 12.5) to be augmented by an area representing *indeterminate stability*. The essential conclusions of this work are shown in Figure 12.6.

The indeterminate-stability region, shown in Figure 12.6, was obtained by relating observations of sustained perturbations in water-fluidized beds to predictions of amplitude growth rates a, eqn (10.3), for a

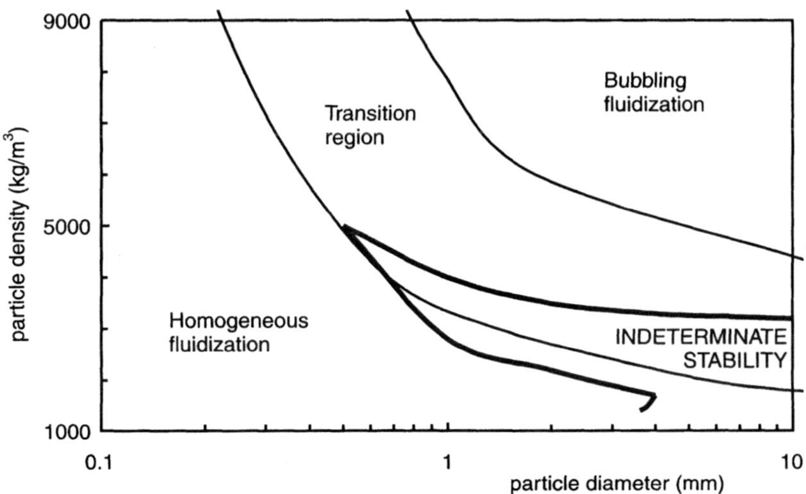

Figure 12.6 Indeterminate-stability region for fluidization by ambient water.

'key wavelength': $\lambda = 20\,d_p$. Systems inside this region are those for which absolute values of a remain below $0.7\,s^{-1}$ over a void fraction range $\Delta\varepsilon$ of at least 0.1. Although the chosen cut-off values for a was decided upon empirically, and involved largely subjective judgements of what in practice constituted persistent, essentially constant amplitude void fraction perturbations, it nevertheless provides a rational, if very approximate, quantification for the observations of these phenomena reported in Chapter 9 and discussed below.

The early studies by Hassett (1961a; 1961b) of inhomogeneities in water-fluidized beds of glass particles revealed heterogeneous behaviour that fell well short of bubbling: low-density, upwards-propagating bands (or *parvoids*), which gradually develop into small, mushroom-shaped voids as the particle diameter is increased beyond about 2 mm. These systems are situated close to the homogeneous-transitional boundary of the global map for fluidization by ambient water, within the indeterminate-stability region of Figure 12.6. The same applies to the systems studied by El-Kaissy and Homsy (1976), where the properties of band-like void fraction waves were measured; this quantitative study has been discussed in Chapter 9 in relation to kinematic-wave propagation through fluidized suspensions.

The minimum bubbling point

Notwithstanding the fact that bubbles in high particle-density, liquid-fluidized beds were reported in the very first comprehensive study of the fluidization process (Wilhelm and Kwauk, 1948), their existence continues to be regarded as something of an anomaly. In gas-fluidized systems, the transition from homogeneous to bubbling fluidization is abrupt and distinct; there is no ambiguity regarding the minimum bubbling point, which clearly separates a very stable state from a highly disturbed one. In liquid systems, for reasons described above, this is not usually the case: the transition is more likely to be gradual, taking place over a range of void fraction, with no clearly identifiable value for ε_{mb}. These remarks apply to the more commonly encountered liquid beds: typically water-fluidized particles of moderate density – up to, say, $3000\,kg/m^3$. High particle-density liquid systems (water fluidization of copper and lead particles, for example) display closer similarities with gas fluidization, with well-defined minimum bubbling points: it is only for these systems that direct comparisons of measured ε_{mb} values with the

Table 12.2 Stability of liquid-fluidized beds: comparison of the experimental observations of Harrison *et al.* (1961) with model predictions

System		Re_t	*Observed behaviour*	*Model predictions*
Lead fluidized by glycerol–water mixtures		2.1	Homogeneous	Always homogeneous
		6.3	Homogeneous	Always homogeneous
		54.3	Bubbling	$\varepsilon_{mb} = 0.56$
		316.5	Bubbling	$\varepsilon_{mb} = 0.40$
Fluidization by paraffin	Resin	8.7	Homogeneous	Always homogeneous
	Glass	32.5	Homogeneous	Always homogeneous
	Steel	72.1	Bubbling	$\varepsilon_{mb} = 0.50$
	Lead	100.0	Bubbling	$\varepsilon_{mb} = 0.42$

model predictions are reported below. For the rest it is only possible to draw broader comparisons between theory and experiment.

An early, comprehensive investigation into liquid-fluidized bed stability by Harrison *et al.* (1961) leads to immediate comparisons with the theory. These experiments involved lead particles ($d_p = 0.77$ mm, $\rho_p = 11\,329$ kg/m^3) fluidized in four different glycerol–water mixtures, and four different particle species (resin, glass, steel and lead), each fluidized by paraffin ($\rho_f = 780$ kg/m^3, $\mu_f = 0.002$ Ns/m^2) – eight systems in all. The experiments consisted of simply observing the nature of the fluidization that occurred in each case: these were reported as being either *stable* (homogeneous) or else *vigorously agitated* (bubbling). Although minimum bubbling points were not reported, it will be seen from Table 12.2 that the observations are all in full accord with the model predictions: systems found to be homogeneous are predicted to be always homogeneous, and those observed to bubble correspond to predicted ε_{mb} values ranging from 0.4 to 0.56.

Water fluidization of copper particles

A systematic study of stability in high solid density, water-fluidized systems is reported by Gibilaro *et al.* (1986). Sieve cuts of copper particles ($\rho_p = 8710$ kg/m^3) were fluidized with water at temperatures ranging from 10 °C to 50 °C. These systems all displayed clear minimum bubbling points. The steady-state characteristics for the homogeneous expansion regions were also reported, enabling measured u_t and n values (which

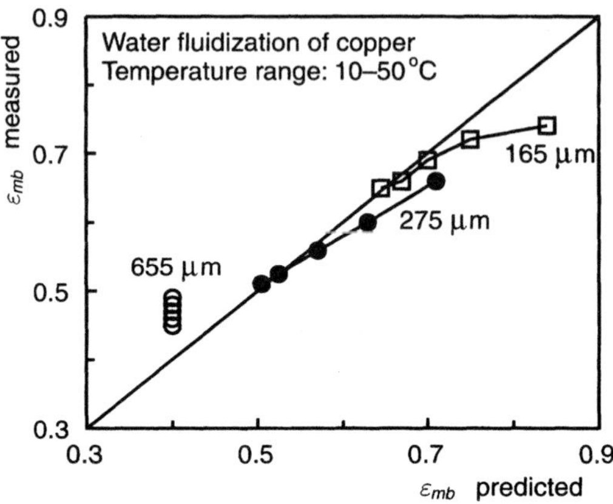

Figure 12.7 Water-fluidization of copper, for temperatures ranging from 10–50 °C: comparison of ε_{mb} values measured by Gibilaro *et al.* (1986) with model predictions.

differed somewhat from those obtained from the standard correlations) to be employed in the predictions of ε_{mb}: these are compared with measured values in Figure 12.7. The data all followed predicted trends with both particle size and temperature. Lower solid density systems (water fluidization of glass and zirconia particles) were also included in this study. These exhibited the propagating band behaviour discussed above, in broad agreement with the model predictions.

Conclusions

Although there is far less experimental data available on the stability of liquid-fluidized beds than there is for gas beds, the results presented above provide further reassuring evidence for the basic integrity of the particle bed model formulation. In some respects the liquid system comparisons go further than those for gas systems, in that they allow for the interpretation of instabilities that fall well short of bubbling behaviour, relating this phenomenon to model predictions of growth- and decay-rates of void fraction perturbations.

The applications of the model have so far involved the linearized forms of the defining equations. In the following chapters the full, non-linear formulation is employed in uncovering further features of the fluidized state that are likewise amenable to predictive verification.

References

El-Kaissy, M.M. and Homsy, G.M. (1976). Instability waves and the origin of bubbles in fluidized beds. *Int. J. Multiphase Flow*, **2**, 379.

Gibilaro, L.G., Hossain, I. and Foscolo, P.U. (1986). Aggregate behaviour of liquid-fluidized beds. *Can. J. Chem. Eng.*, **64**, 931.

Gibilaro, L.G., Di Felice, R., Foscolo, P.U. and Waldram, S.P. (1988). Fluidization quality: a criterion for indeterminate stability. *Chem. Eng. J.*, **37**, 25.

Harrison, D., Davidson, J.F. and de Kock, J.W. (1961). On the nature of aggregative and particulate fluidization. *Trans. Inst. Chem. Eng.*, **39**, 202.

Hassett, N.L. (1961a). Flow patterns in particle beds. *Nature*, **189**, 997.

Hassett, N.L. (1961b). The mechanism of fluidization. *Br. Chem. Eng.*, **11**, 777.

Wilhelm, R.H. and Kwauk, M. (1948). Fluidization of solid particles. *Chem. Eng. Prog.*, **44**, 201.

13

The scaling relations

Cold-model simulations

Fluidized beds are used in a wide spectrum of large-scale process applications, often involving high operating temperatures and pressures. The construction and commissioning of such equipment is extremely costly, so that any uncertainty at the design stage regarding the fluidization quality that will result in the completed plant represents a major cause for concern. Laboratory bench-scale experiments, which may well have been employed to test the feasibility of the basic process, are of limited help here, as the size of bubbles in the small scale prototype (to take just one important fluidization quality parameter) provides little indication of what this will be in the commercial unit. Some reassurance could possibly be provided by the fluidization quality map presented in Chapter 10, perhaps relating the proposed plant to a system for which the fluidization quality is well documented. However, this may prove insufficient, particularly if the bed is to contain heat exchanger tubes or other internals, which modify the fluid flow field, rendering it significantly different to its otherwise matched partner.

One way of tackling the problem is to build a model bed in which the fluidization quality of the proposed plant can be simulated and studied. Only the fluidization characteristics need be considered, so that the model may be operated without the heat transfer and chemical reaction processes required of the envisaged commercial unit; it may therefore be operated under ambient conditions of temperature and pressure (or perhaps under somewhat elevated pressure) and so be constructed cheaply, perhaps using transparent material through which the behaviour may be directly observed. The particles, fluid and operating conditions must be chosen so as to ensure equivalence of the *cold model* to the final plant: it is the *scaling relations* that provide the criteria for making these choices.

The dimensionless equations of change

A necessary condition for a cold model to simulate the fluidization characteristics of an envisaged plant is that the defining equations for the two units, and the *numerical values of the parameters* in those equations, be identical. This condition can generally be satisfied by first expressing the equations in dimensionless form; the resulting parameters then represent dimensionless combinations of those of the physical system, thereby providing for some flexibility in matching numerical values. This technique, together with the more fundamental dimensional analysis method of Rayleigh and the Π method of Buckingham (see Massey, 1971), has a long and distinguished history in the study of single-phase fluid systems, which, by way of introduction to fluidized bed applications, we now briefly consider.

Single fluid systems

The x-direction component of the Navier-Stokes equation for momentum conservation in a Newtonian fluid of constant density and viscosity is given by:

$$\rho_f\left(\frac{\partial v_x}{\partial t}+v_x\frac{\partial v_x}{\partial x}\right)=-\frac{\partial p}{\partial x}+\mu_f\left(\frac{\partial^2 v_x}{\partial x^2}+\frac{\partial^2 v_x}{\partial y^2}+\frac{\partial^2 v_x}{\partial z^2}\right)+\rho_f g_x, \quad (13.1)$$

where ρ_f and μ_f are the fluid density and viscosity, v_x the fluid velocity in the x direction, p the pressure, and g_x the x-component of the gravitational field strength.

To render eqn (13.1) dimensionless, it is first necessary to select convenient reference levels for all the variables: x, y, z, t, v_x and p. The choice is quite arbitrary. For the distance variables (x, y and z) the reference level L would typically consist of a key equipment dimension – a tube or tank diameter, or the length of a submerged object, etc.; for velocity v_x, an average value or an entering volumetric flux V; and for the remaining variables (t and p) appropriate combinations of the other reference level can be constructed: for example, L/V for t, and $\rho_f V^2$ for p.

Rewriting eqn (13.1) in terms of the dimensionless variables, $\hat{x}$, $\hat{y}$, $\hat{z}$, $\hat{t}$, $\hat{v}_x$ and $\hat{p}$, defined by: $x = \hat{x} \cdot L$, $t = \hat{t} \cdot L/V$, etc., we obtain:

$$\frac{\partial \hat{v}_x}{\partial \hat{t}} + \hat{v}_x \frac{\partial \hat{v}_x}{\partial \hat{x}} = -\frac{\partial \hat{p}}{\partial \hat{x}} + \left(\frac{1}{Re}\right) \cdot \left(\frac{\partial^2 \hat{v}_x}{\partial \hat{x}^2} + \frac{\partial^2 \hat{v}_x}{\partial \hat{y}^2} + \frac{\partial^2 \hat{v}_x}{\partial \hat{z}^2}\right) + \left(\frac{1}{Fr}\right), \tag{13.2}$$

which contains just two dimensionless parameters, the Reynolds number Re and the Froud number Fr:

$$Re = \frac{LV\rho_f}{\mu_f}, \qquad Fr = \frac{V^2}{Lg_x}. \tag{13.3}$$

Many physical systems may be constructed and operated in a manner that results in Re and Fr having the same numerical values; such systems are referred to as being *dynamically similar*. If, in addition, the dimensionless boundary conditions are the same, which is usually the case if the systems are *geometrically similar* (that is to say, one represents a scale model of the other), then the flow behaviour of matched systems, expressed in terms of the dimensional variables, will be identical. This has provided the basis for countless cold-modelling studies, firmly establishing the procedure at the forefront of experimental process research.

The scaling relations for fluidization

Given the long, successful record of scaling experimentation in the study of complex flow behaviour of fluids, it is somewhat surprising that it was not until the mid-1980s that scaling relations for fluidization were applied to cold-modelling studies of fluidization quality (Fitzgerald *et al.*, 1984; Glicksman, 1984). These relations emerge from the equations of change on following exactly the procedure illustrated above for the case of a single fluid. They will now be derived from the *particle bed* model equations (Foscolo *et al.*, 1990, 1991).

We start with the one-dimensional, two-phase formulation reported in Chapter 8: eqns (8.21)–(8.24), together with the constitutive relations, eqns (8.4) and (8.18). The primary reference levels may be chosen with regard to particle characteristics: d_p for distance and u_t for velocity. On this basis, all the variables may be related to their dimensionless counterparts:

$$z = \hat{z} \cdot d_p, \quad t = \hat{t} \cdot d_p/u_t, \quad u_f = \hat{u}_f \cdot u_t, \quad p = \hat{p} \cdot \rho_p u_t^2. \tag{13.4}$$

Substitution of these variables in the equations of change yields:

Conservation of mass

$$\frac{\partial \varepsilon}{\partial \hat{t}} + \frac{\partial}{\partial \hat{z}}\left[\varepsilon \hat{u}_f\right] = 0, \quad \textit{Fluid phase} \tag{13.5}$$

$$\frac{\partial \varepsilon}{\partial \hat{t}} - \frac{\partial}{\partial \hat{z}}\left[(1-\varepsilon)\hat{u}_f\right] = 0, \quad \textit{Particle phase} \tag{13.6}$$

Conservation of momentum

$$De\left[\frac{\partial \hat{u}_f}{\partial \hat{t}} + \hat{u}_f \frac{\partial \hat{u}_f}{\partial \hat{z}}\right] + \frac{De}{Fr} + (1-\varepsilon)\frac{(1-De)}{Fr}(Fl - \hat{u}_p)^{4.8/n}\varepsilon^{-4.8} + \frac{\partial \hat{p}}{\partial \hat{z}} = 0, \quad \textit{Fluid phase} \tag{13.7}$$

$$\frac{\partial \hat{u}_p}{\partial \hat{t}} + \hat{u}_p \frac{\partial \hat{u}_p}{\partial \hat{z}} + \frac{1}{Fr} - \frac{(1-De)}{Fr}(Fl - \hat{u}_p)^{4.8/n}\varepsilon^{-3.8} - 3.2\frac{(1-De)}{Fr}\frac{\partial \varepsilon}{\partial \hat{z}} + \frac{\partial \hat{p}}{\partial \hat{z}} = 0. \quad \textit{Particle phase} \tag{13.8}$$

The above formulation is in terms of four dimensionless parameters, the *density number De*, the *Froud number Fr*, the *Reynolds number* Re_t (which appears implicitly through the Richardson–Zaki exponent n, which depends solely on Re_t ($n = n(Re_t)$), and the *Flow number Fl*:

$$De = \frac{\rho_f}{\rho_p}, \quad Fr = \frac{u_t^2}{g d_p}, \quad Re_t = \frac{\rho_f d_p u_t}{\mu_f}, \quad Fl = \frac{U_0}{u_t}. \tag{13.9}$$

This group of four controlling parameters may be reduced to three and expressed somewhat more conveniently. The reduction in number comes about as a result of selecting d_p and u_t as the primary reference levels. The force balance for a single, unhindered particle may be written:

$$\frac{\pi d_p^3}{6}(\rho_p - \rho_f)g = C_D \frac{\rho_f u_t^2}{2} \cdot \frac{\pi d_p^2}{4}, \tag{13.10}$$

where the drag coefficient C_D depends solely on Re_t. Equation (13.10) yields the relation between C_D (and hence Re_t) and two of the above dimensionless groups:

$$C_D = \frac{4(1 - De)}{3FrDe}, \tag{13.11}$$

so that Re_t becomes redundant as a controlling parameter. The Froud and Reynolds numbers may be combined to eliminate u_t and produce the *Galileo number Ga*, which may then replace *Fr*:

$$Ga = \frac{Re_t^2}{Fr} = \frac{g d_p^3 \rho_p^2}{\mu_f^2}. \tag{13.12}$$

The conditions for one-dimensional similarity may thus be expressed in terms of just three dimensionless groups.

The one-dimensional scaling parameters:

$$Ga = \frac{g d_p^3 \rho_f^2}{\mu_f^2}, \quad De = \frac{\rho_f}{\rho_p}, \quad Fl = \frac{U_0}{u_t}. \tag{13.13}$$

The *Archimedes number Ar*, introduced in earlier chapters, is closely related to *Ga* and may be used in its place:

$$Ar = Ga(1 - De)/De. \tag{13.14}$$

Experimental verification of the one-dimensional scaling rules

The major incentive for developing scaling rules is to enable complex three-dimensional phenomena, relating to large-scale equipment, to be studied experimentally at relatively low cost by means of smaller-scale 'cold models'. The above one-dimensional rules, which do not include a geometric similarity parameter, are therefore of limited applicability, but may nevertheless be applied to the essentially one-dimensional problem of predicting the onset of unstable, bubbling behaviour. As will now be demonstrated, the rules enable quite different physical systems to be matched with regard to this dramatic event, thereby offering grounds for confidence in more general applications.

Table 13.1 shows three pairs of fluidized systems, each representing a water- and a gas-fluidized bed, which are approximately matched with regard to the one-dimensional scaling rules. The results for fluidization with synthesis gas at 124 bar (Jacob and Weimer, 1987) and with water at 10 °C (Gibilaro *et al.*, 1986) have been referred to in Chapters 9 and 12 respectively, where they are shown to support the predictions of the *particle bed* model for the minimum bubbling void fraction ε_{mb}; the high pressure carbon tetrafluoride results (at 21 bar and 69 bar for systems I and III respectively) were reported by Crowther and Whitehead (1978).

The System-I pair, water-fluidized copper and gas-fluidized alumina, are closely matched with regard to the scaling parameters, and both start to bubble at approximately the same void fraction: $\varepsilon_{mb} = 0.66$ and 0.68 respectively. In neither case is there any bed contraction at the minimum bubbling point, as is usual in liquid systems, but, for reasons discussed in Chapter 14, only to be observed in gas systems under very high pressure conditions. The System-II pair, water-fluidized copper and high-pressure, synthesis-gas-fluidized fine carbon, are also reasonably well matched. Both manifest extensive regions of homogeneous expansion, up to void fractions of 0.74 and 0.80 respectively. Once again, neither system exhibits any contraction at the minimum bubbling point, the bed height/void

Table 13.1 Near dynamic similarity of matched gas- and water-fluidized systems

	Fluidized system					*Dynamic similarity*			
		ρ_f kg/m^3	μ_f Ns/m^2 $\times 10^5$	d_p μm	ρ_p kg/m^3	*De*	*Ga*	*Fl* at ε_{mb}	*Stability*
I	water/copper	1000	130	275	8700	0.12	121	0.24	$\varepsilon_{mb} = 0.66$
	CF_4/alumina	120	1.8	63	900	0.13	109	0.20	$\varepsilon_{mb} = 0.68$
II	water/copper	1000	130	165	8700	0.12	26	0.27	$\varepsilon_{mb} = 0.74$
	synthesis-gas/carbon	82.5	1.6	44	850	0.10	22	0.36	$\varepsilon_{mb} = 0.80$
III	Water/soda-glass	1000	100	400	2500	0.40	628	–	stable
	CF_4/alumina	388	2.35	63	900	0.43	669	–	stable

fraction relation merely displaying a reduction of gradient in the bubbling region beyond ε_{mb} – as was also the case for the first matched pair. The System-III pair provides a comparison of a commonly reported water-fluidized bed (0.4 mm soda-glass) with the high-pressure gas-fluidization of alumina: in both cases the stable, homogeneous state is preserved over the entire expansion range.

A generalized powder classification for fluidization by any fluid

In the opening section of Chapter 9 a powder classification map for fluidization by ambient air was constructed on the basis of predictions of the *particle bed* model. This represented a theoretical counterpart of much of the empirical Geldart classification, defining regions for which the fluidization is always bubbling (group B), always homogeneous or cohesive (group C), or displays a transition from homogeneous to bubbling behaviour (group A). The group D region was added to the theoretical map in Chapter 10, completing the correspondence of the empirical and theoretical classifications. This followed the discovery that particles falling into this group give rise to very low, physically unobtainable, ε_{mb} predictions of less than 0.1.

We are now in a position to generalize the theoretical map so as to encompass fluidization by any fluid. The construction procedure described in the opening section of Chapter 9 may be readily adapted for this purpose, the reciprocal of the density number De^{-1} replacing particle density ρ_p, and the Galileo number Ga replacing particle diameter d_p. Thus, for selected values of De^{-1}, values of Ga corresponding to positions on the region boundaries are obtained by iteration to satisfy the various conditions for the stability function $S(\varepsilon)$, eqn (9.1), expressed in terms of the dimensionless groups selected above:

$$S = \frac{1.79\varepsilon^{1-n}}{n} \cdot \sqrt{\frac{Fr(1 - De)}{1 - \varepsilon}} - 1 = \begin{cases} +\text{ve: homogeneous} \\ 0\text{: stability limit} \\ -\text{ve: bubbling} \end{cases} \qquad (13.15)$$

The procedure is as follows. For specified values of De and Ga, the Archimedes number Ar is first calculated from eqn (13.14); then the Reynolds number Re_t from eqn (2.17); then n from eqn (4.5); then the Froud number Fr from eqn (13.12): $Fr = Re^2/\text{Ga}$; and finally S from eqn (13.15). The region boundaries are then simply the combinations of De^{-1} and Ga that satisfy the following conditions:

CA boundary: $S = dS/d\varepsilon = 0.$
AB boundary: $S(\varepsilon = 0.4) = 0.$
BD boundary: $S(\varepsilon = 0.1) = 0.$

The resulting fluidization map is shown in Figure 13.1. Regions typically encountered for ambient air and ambient water systems are also indicated. As we have seen in previous chapters, the model predictions in these regions are in good accord with empirical observations. High-pressure gas-fluidized systems fall below the ambient air region, approaching that for ambient water fluidization; empirical observations of the AB

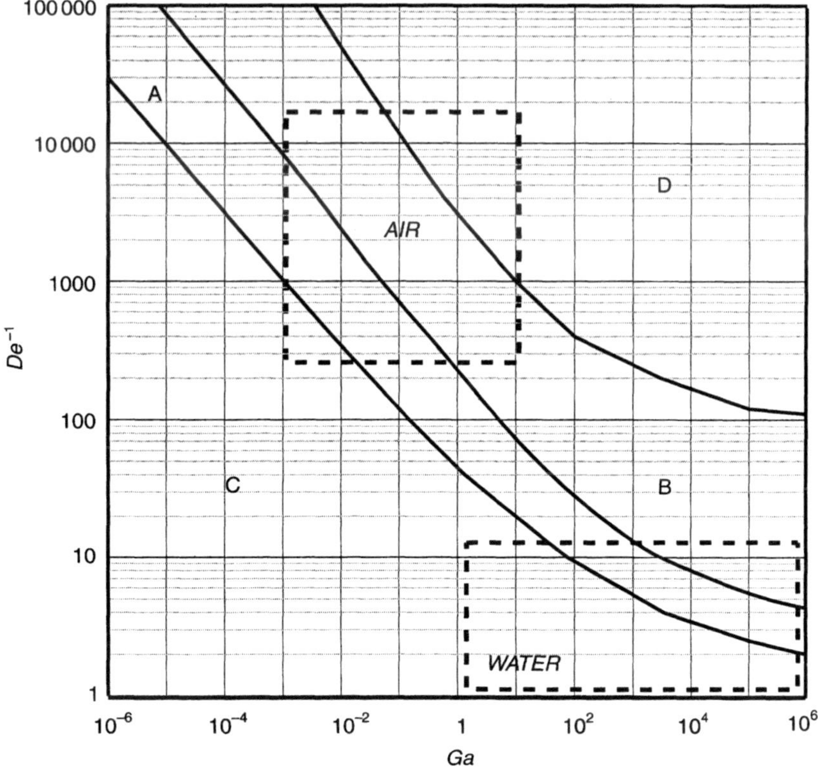

Figure 13.1 Generalized powder classification for fluidization by any fluid – showing the Geldart classification boundaries (A, B, C and D) and regions corresponding to ambient air and water fluidization.

boundary for such cases have been reported by Grace (1986), in reasonable agreement with the predicted trend (Foscolo *et al.*, 1991).

The three-dimensional scaling relations for geometrically similar fluidized beds

Up to now we have considered only the axial flow direction. The generalization of the *particle bed* model equations for multidirectional flow is considered in Chapter 16, where it will be seen to replicate the development of bubbles and other inhomogeneities in unstable systems by numerical simulation. For present purposes it is only necessary to point out that the defining equations for the lateral flow directions add no further dimensionless groups to those obtained above for axial flow alone. It is only the boundary conditions that impose further similarity criteria, but for *geometrically similar* systems these reduce in practice to simply matching the *length number Le*: $Le = L/d_p$, where L is some representative length dimension. In principle, the dimensionless pressure boundary condition, $p_0/\rho_p u_t^2$, should also be matched, but this has been shown to be unnecessary for all cases of practical interest (Glicksman, 1984). Significant particle size distributions should also be matched, as well as particle shape for non-spherical particle systems.

The three-dimensional scaling parameters:

$$Ga = \frac{g d_p^3 \rho_f^2}{\mu_f^2}, \quad De = \frac{\rho_f}{\rho_p}, \quad Fl = \frac{U_0}{u_t}, \quad Le = \frac{L}{d_p}. \tag{13.16}$$

Selection of cold-model parameters by means of the above dimensionless groups is very simple. Once the fluidizing gas and convenient conditions (for example, ambient) have been chosen, the particle density is fixed by the density number *De*, after which its diameter follows from the Galileo number *Ga*. The length number *Le* and flow number *Fl* then dictate the size of the scale model and the operating fluid flux respectively:

$$\text{Cold-model } \textit{scale-factor} = d_p(\text{model})/d_p(\text{system}),$$
$$\text{Cold-model } \textit{flux-factor} = u_t(\text{model})/u_t(\text{system}).$$

The expression for *Ga* indicates that in order to reduce the diameter of the test particle and, as a consequence, the size of the cold model, the fluid density should be increased and its viscosity decreased; for gas fluidiza-

Table 13.2 Cold models for a high temperature fluidized bed catalytic reactor

		Catalytic reactor	*Cold models*	
			Ambient pressure	*Elevated pressure*
Gas properties	Material	Air	Air	Air
	Temperature, °C	800	20	20
	Pressure, bar	1.0	1.0	2.5
	Density, kg/m^3	0.33	1.21	3.03
	Viscosity, $Ns/m^2 \times 10^5$	4.4	1.8	1.8
Particle properties	Material	Alumina	Zirconia	Copper
	Diameter, mm	1.0	0.23	0.13
	Density, kg/m^3	1000	3667	9182
u_t	m/s	4.31	2.06	1.59
Scaling parameters	*De*	0.00033	0.00033	0.00033
	Ga	0.55	0.54	0.61
Flux factor			0.48	0.36
Scale factor			0.23	0.13

tion both of these effects can be achieved by lowering the operating temperature.

Table 13.2 provides illustrations of how the fluid-dynamic behaviour of hot alumina particles, fluidized at atmospheric pressure, can be studied in smaller ambient temperature beds using the same fluid (air in the chosen example). The first cold model, operating under the same ambient pressure condition as the high temperature reactor, achieves a scale factor of 0.23; this may be further reduced by operating at a somewhat elevated pressure, as illustrated by the second cold model.

For cases where the industrial process involves high temperatures and pressures, as in pressurized fluidized-bed combustion, it becomes possible to select the pressure in the cold model in such a way as to maintain the gas density, and hence the particle material, the same as in the industrial unit. This is illustrated in the second cold model of Table 13.3, where, for the chosen example, it results in a scale factor of approximately one-half. The first cold model of Table 13.3 operates with the same gas as the industrial unit under conditions of ambient pressure as well as ambient temperature. The fact that the model size is only marginally less than the original does not necessarily impose a severe limitation, as the effect of bed internals

Table 13.3 Cold models for a pressurized fluidized bed combustor

		Fluidized combustor	*Cold models*	
			Ambient pressure	*Same fluid and particle densities*
Gas properties	Material	Air	Air	Air
	Temperature, °C	750	20	20
	Pressure, bar	7.0	1.0	2.0
	Density, kg/m^3	2.43	1.21	2.43
	Viscosity, $Ns/m^2 \times 10^5$	4.3	1.8	1.8
Particle properties	Material	Silica	Alumina	Silica
	Diameter, mm	1.0	0.89	0.56
	Density, kg/m^3	2500	1240	2500
u_t	m/s	3.88	3.65	2.91
Scaling parameters	*De*	0.00097	0.00098	0.00097
	Ga	31.33	31.25	31.40
Flux factor			0.94	0.75
Scale factor			0.89	0.56

(heat exchanger tubes, baffles, etc.) is often to partition the industrial bed into smaller cells, which can be modelled independently.

Compatibility of the scaling relations with other formulations

The relations presented in this chapter, based on the *particle bed* model formulation of the defining equations for fluidization, are fully compatible with those previously derived by other workers, including Glicksman (1984) and Fitzgerald *et al.* (1984), which were based on the original formulation by Jackson (1963), which did not include an elasticity term in the particle-phase momentum equation. This means that these other rules could equally well have been used to predict the equivalence of all the matched-system examples considered above, including the one-dimensional cases of Table 13.1, which involved transitions from homogeneous to bubbling behaviour – even though the equations from which these rules were derived exclude the possibility of homogeneous behaviour.

The reason for this paradoxical state of affairs is that particle-phase elasticity in the *particle bed* model formulation is a purely fluid-dynamic phenomenon, dependent on the same process variables that determine the primary fluid-particle interactions. Were this not the case, were particle-phase elasticity a phenomenon that can only exist as a result of non-fluid-dynamic interactions, then another (and in most cases quite unmatchable) scaling parameter would have to be employed. This point has deterred workers from applying the scaling relations to systems capable of fluidizing homogeneously, thereby excluding, quite unnecessarily on the basis of the results presented above, the important industrial area of fine powder fluidization, where homogeneous behaviour at low fluidizing velocities is a manifest reality.

Fluidization quality characterization: fluid pressure fluctuations

When a cold model of a proposed industrial unit has been constructed and operated in accordance with the scaling criteria of eqn (13.16), the next problem is to find some means of evaluating the fluidization quality. Given the complexity of heterogeneous bed behaviour (involving such basic bubble characteristics as size and velocity distribution, coalescence and splitting), it is clear that some practical, indirect measure becomes essential. Fluid pressure fluctuations provide a readily accessible means of characterizing bed heterogeneities. Chapter 16 reports numerical simulations of the two-dimensional *particle bed* model that provide a direct link between fluid pressure fluctuations and bubble-related phenomena. For now the focus will be on experiments designed to test the reproducibility of measured fluid pressure fluctuation characteristics in different, identically scaled systems.

Fluid pressure fluctuation measurement in fluidized beds

Direct pressure measurement

There are various ways in which pressure fluctuation measurements can be obtained. The most straightforward arrangement, utilized in the two experimental programmes reported below for group A and group B powder systems, involves simply transmitting the pressure at a selected bed location through an open tube to a transducer, which outputs via an analogue/digital interface to a computer memory.

With this arrangement the signal fluctuations about their mean value are strongly influenced by bubble eruption at the bed surface, reflecting

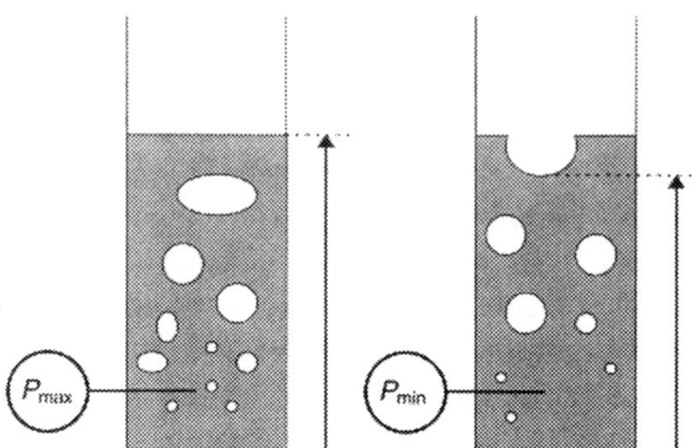

Figure 13.2 The effect of bed surface eruptions on fluid pressure.

maximum bubble size and frequency. Figure 13.2 illustrates the reason for this dependence. Fluctuation amplitudes obtained in this way are relatively insensitive to the location of the pressure probe. This method probably provides the most significant characteristic of fluidization quality for most practical applications.

Differential pressure measurement

Local inhomogeneities may be measured by means of two probes spaced a short axial distance apart in the bed and connected across a differential pressure transducer (Figure 13.3). In contrast to the previously described method, measurements obtained in this way are effectively uninfluenced by bed surface eruptions, reflecting instead the passage of inhomogeneities between the probes. This was the arrangement employed in the *indeterminate stability* study referred to in Chapter 12 (Gibilaro *et al.*, 1988). The probes in that case consisted of 3 mm diameter tubes spaced 10 mm apart.

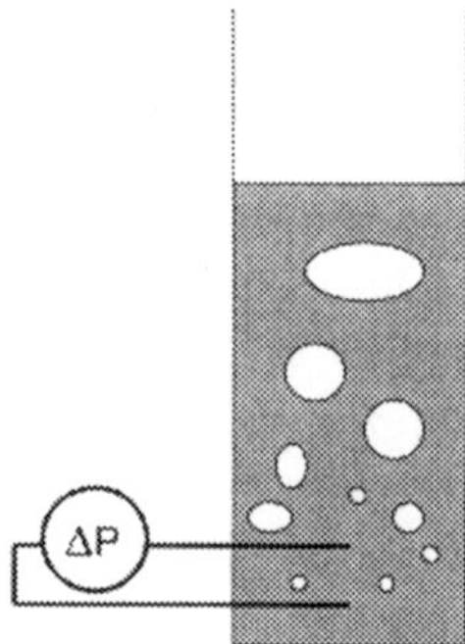

Figure 13.3 Differential pressure measurement.

Two-station measurement

By measuring pressure fluctuations at two locations some distance apart, the velocity of inhomogeneities passing between them can be obtained by cross-correlation (Figure 13.4). This represents a widely used experimental technique for measuring bubble velocities in fluidized beds.

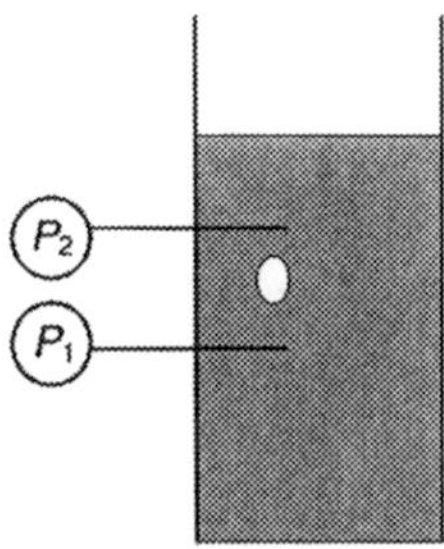

Figure 13.4 Two-station pressure measurement.

Experimental studies of pressure fluctuation in scaled fluidized beds

In order to test the validity of the scaling relations of eqn (13.16), two experimental investigations have been undertaken, the essential findings of which are summarized below. The first involved fine powder (group A) systems that display transitions from homogeneous to bubbling behaviour. It has been reported above that the scaling relations lead to essentially matched values of ε_{mb} in these cases. Fluid pressure fluctuation characteristics in the bubbling regime are reported below. The second investigation involved coarser (group B) powders. These systems are known to conform to previous scaling relations that are completely compatible with those presented here; the reported results, however, draw attention to certain limitations to their applicability.

All experiments were performed with air at ambient temperature as the fluidizing medium. Its density was adjusted where necessary by operating at elevated pressure. This was achieved in a pressure vessel specially designed to enable visual observation of transparent beds through narrow, vertical, perspex windows. Through these the bed surface comportment was observed and recorded on videocassette. The instantaneous pressure, at a fixed level above the distributor plate, was measured continuously with a piezoelectric sensor, and sampled at a frequency of 20 Hz

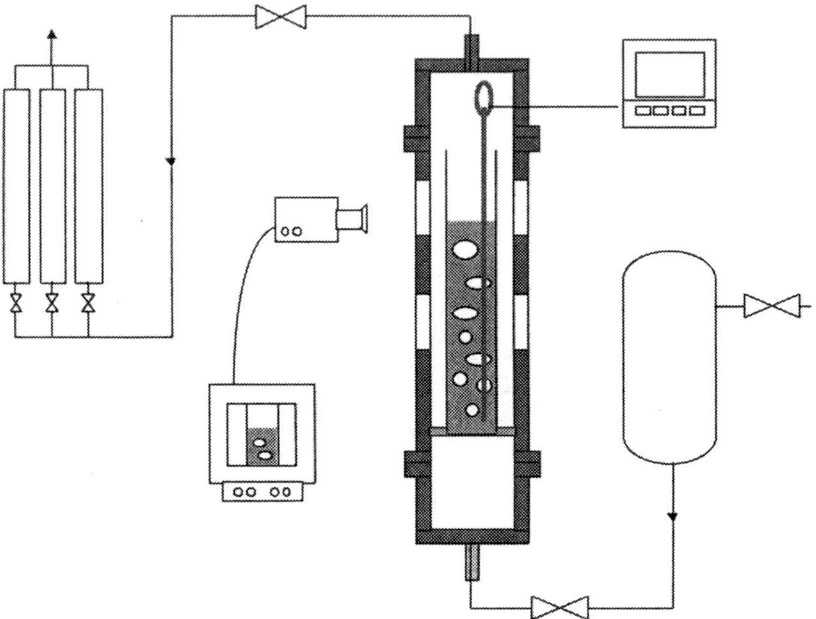

Figure 13.5 Equipment for observing fluidized bed behaviour under high pressure conditions.

over 10 s time intervals. The experimental rig, described in more detail in Foscolo *et al.* (1989), is illustrated in Figure 13.5.

Fine powder (Geldart group A) fluidization experiments

Results have been published (Rapagnà *et al.*, 1992) for five pairs of fine powder fluidized beds, matched in accordance with the scaling relations of eqn (13.16) and tested as described above. Fluid pressure fluctuations were logged over extensive ranges of air flux, starting from just in excess of minimum bubbling values U_{mb}. The pressure data were processed using standard fast Fourier transform software to produce the frequency power spectrum and the root mean square $\widehat{p}$ of the fluctuations. The scaling requirements dictated wide variations in system properties: particle densities ranging from 900 to 9000 kg/m^3, air pressures from 0.9 to 10 bar, particle diameters from 14 to 100 μm, bed diameters from 50 to 200 mm.

Particle species (soda-glass, copper and diverse cracking catalyst supports) varied widely with regard to material properties (porosity, electrical conductivity, Hamaker constant, etc.), so that any influence of

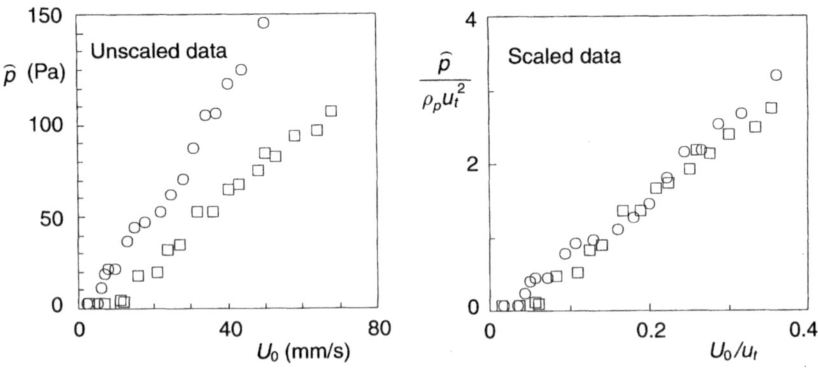

Figure 13.6 Pressure fluctuations in scaled, fine powder fluidized beds: RMS pressure vs. fluid flux for typical example of results reported by Rapagnà *et al.* (1992).
Bed 1 (circles): air/soda-glass, $p = 2.2$ bar, $d_p = 47\,\mu$m, $\rho_p = 2540\,\text{kg/m}^3$.
Bed 2 (squares): air/catalyst, $p = 0.92$ bar, $d_p = 86\,\mu$m, $\rho_p = 1054\,\text{kg/m}^3$.
Approximate values of scaling parameters: $Ga = 0.022$, $De = 0.001$, $Le = 1150$.

particle–particle force interaction would have the effect of destroying dynamic similarity; in this way, the experiments were effectively tailored to differentiate between fluid-dynamic and interparticle-force explanations for the initial region of homogeneous fluidization. In the event, excellent agreement was found for all aspects of scaled behaviour for each of the five pairs of matched beds, including minimum bubbling points (ε_{mb}, U_{mb}) and scaled pressure fluctuation characteristics; these latter exhibited broadly equivalent dimensionless frequency bands and closely matched dimensionless root mean square values over the full range of operation.

Figure 13.6 displays root mean square evaluations $\hat{p}$ as functions of fluid flux for an example matched pair that is typical in this respect of all the five tested. The left-hand figure shows the trends of *unscaled* date points (U_0, $\hat{p}$), the two systems diverging progressively with increasing fluid flux. On the other hand, the *scaled* results on the right (U_0/u_t, $\hat{p}/\rho_p u_t^2$) are in excellent agreement over the entire operating range.

Coarse powder (Geldart group B) fluidization experiments

A similar programme to the one just described has been conducted on beds of coarser particles, which bubble from the onset of fluidization

Table 13.4 Fluidization of group B powders by ambient temperature air

	Scaled systems			*Unscaled systems*	
	Lapasorb □	*Sand* △	*Bronze* ○	*Iron* ▲	*Sand* ◆
p (bar)	0.92	2.0	6.6	6.2	0.92
ρ_p (kg/m^3)	1216	2640	8770	7300	2640
d_p (μm)	597	348	158	163	348
D (mm)	192	106	49.5	49.5	192
u_t (m/s)	2.62	1.98	1.34	1.25	2.59
Ga	7.79	7.35	7.64	7.38	1.54
De ($\times 10^4$)	9.05	9.09	9.12	10.3	4.2

(Di Felice *et al.*, 1992). This time tests were performed on three systems, each scaled for equivalence in accord with the relations of eqn (13.16). Air at ambient temperature was the fluidizing medium in every case, its pressure being adjusted, as before, to suit the scaling requirements. The particles were essentially spherical, and their size distributions were very similar. Two further systems were also tested, both of which violated the scaling requirements in some way. One involved angular iron particles, which satisfied the scaling criteria reasonably closely in all respects other than their non-spherical geometry. The other, involving the same sand particles as one of the scaled systems, was fluidized under conditions that put it way off scale. Two different aspect ratios (bed height H divided by bed diameter D) were considered. For reasons that will soon become apparent, this factor can be crucial in setting limits to the applicability of fluid-dynamic scaling. Table 13.4 summarizes the system properties.

Experiments at low aspect ratio – $H/D \approx 2.8$

An important feature of group B system behaviour is that the bubbles grow, largely by coalescence, as they rise through the bed. This represents a major cause for concern in the design of fluidized bed reactors, prompting the adoption of cold-model studies, because large bubbles can mean poor gas–solid contact, and hence a reduction in chemical conversion. The higher the aspect ratio of the bed, the greater the opportunity for bubbles to grow, perhaps enabling their diameter to approach that of the bed itself. If this occurs, then bubbling fluidization gives way to *slugging*,

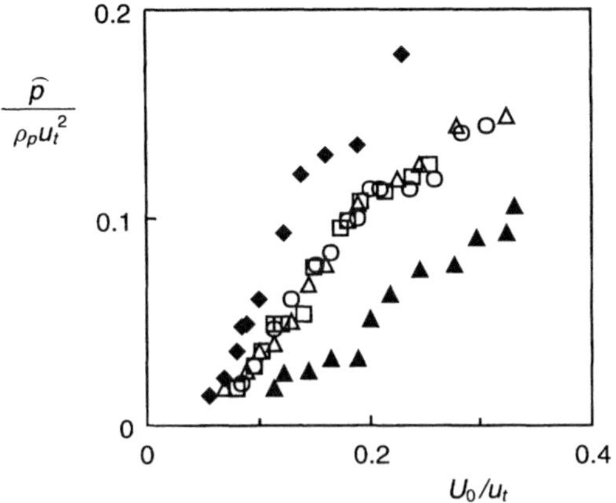

Figure 13.7 Bubbling fluidization in *shallow* beds; root mean square of dimensionless pressure fluctuations vs dimensionless fluid flux (symbols as in Table 13.4).

and a very different behaviour pattern ensues. Some interesting aspects of slugging behaviour are analysed in Chapter 15. For now it is sufficient to report that for the more shallow bed condition adopted in the study, $H/D \approx 2.8$, bubbling conditions prevailed.

The behaviour of the three scaled beds of Table 13.4 turned out to be closely matched when compared in terms of dimensionless variables; minimum fluidization flux, bed expansion and bubble holdup all conformed in this respect. The dimensionless pressure fluctuations were also very similar, the best measure being provided by the dimensionless root mean square, $\widehat{p}/\rho_p u_t^2$. This is shown in Figure 13.7 as a function of dimensionless fluid flux, U_0/u_t: the three matched systems (open symbols) form an essentially single curve sandwiched between those of the two unmatched systems (solid symbols).

Experiments at high aspect ratio – $H/D \approx 5.4$

All the tests carried out on the relatively shallow beds were repeated at approximately double the aspect ratio. This resulted in predominantly slugging behaviour in all systems, the bubble sizes reaching bed dimensions to form *gas-slugs* which, interspersed with piston-like *solid slugs*, travel upwards through the bed, giving rise to regular, large-amplitude surface oscillations. Two very clear conclusions concerning the

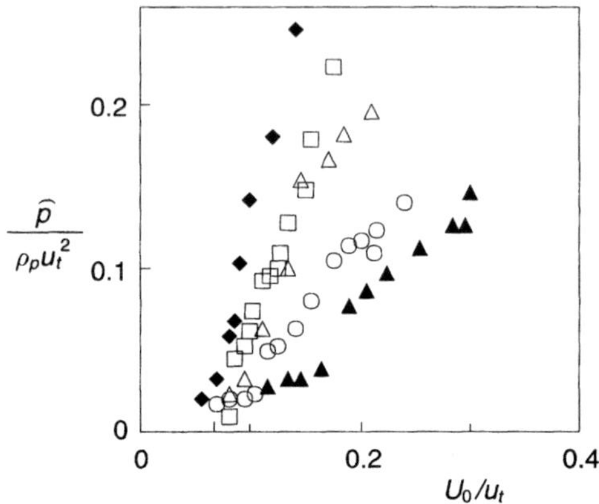

Figure 13.8 Slugging fluidization in *deep* beds: root mean square of dimensionless pressure fluctuations vs dimensionless fluid flux (symbols as in Table 13.4).

pressure fluctuation characteristics emerge from these experiments; the first, a rather obvious one, is that they become periodic in nature, reflecting the cyclic rise and collapse of the bed surface. This aspect is also reflected in the frequency power spectrum, which displays a narrow range of clearly dominant frequencies – unlike the case for low aspect ratio beds.

The second and more pertinent observation is that dimensionless equivalence of the three scaled beds, so amply confirmed in the low aspect ratio experiments, is completely destroyed. This is well illustrated by comparison of Figure 13.8 with its counterpart for shallow beds, Figure 13.7; the scaled systems no longer compact to a single curve, but spread out over the region bounded by the two unscaled systems. This points to the presence of non-fluid-dynamic interactions, which will be discussed in Chapter 15: fluid-dynamically based scaling alone is insufficient for cold-modelling studies of slugging systems.

This last conclusion is somewhat worrying, as slugging behaviour is probably more common in industrial plant than is commonly appreciated. This point has been made by Grace and Harrison (1970), and has to do with the effect of bed internals alluded to earlier: vertical heat-exchanger tubes, for example, can effectively subdivide a bed into smaller diameter, high aspect ratio cells. The advantage this poses in terms of reducing the size of the cold model could be offset by the uncertain validity of the scaling relations.

Comparison of the scaling relations with the fluidization quality criteria

The scaling relations considered in this chapter and the fluidization quality criteria derived in Chapter 10 are both based on the same fluid-dynamic description of the fluidized state. However, whereas the scaling relations guarantee full fluid-dynamic equivalence in matched systems, the fluidization quality parameters, ε_{mb} and Δa, relate only to the expected degree of instability, reflecting such things as bubble size, velocity and frequency. This deficiency is compensated for by an increase in flexibility, which can be appreciated very easily from the following argument.

Consider the situation of differing particle species fluidized by a given fluid. It is well known empirically that the extent of instability increases with both particle size and density; hence an increase in particle size d_p can be compensated for in this respect by a decrease in particle density ρ_p. This manoeuvre, however, is completely at odds with the scaling rules: d_p appears in the Galileo number Ga, and ρ_p in the density number De; so that we are here correcting for an imbalance in one dimensionless group by creating a further imbalance in another. The fluidization quality criteria, on the other hand, have no difficulty with such a procedure. Moreover, the fact that they fail to match other conditions, such as the minimum fluidization velocity, is of little disadvantage in practice, as in general these are readily available from independent correlations: it is fluidization quality that is the main cause for concern in new proposed applications.

The reason that the fluidization quality approach can lead to matching conditions which do not satisfy scaling requirements has to do with the fact that the parameter Δa has the dimensions of reciprocal time. Were it to be expressed in dimensionless form, as is the case for the other parameter ε_{mb}, then fluidization quality matching would give identical results to those obtained from the scaling relations. This can be readily demonstrated as follows.

Fluidization quality was characterized in Chapter 10 by means of the parameters ε_{mb} and Δa, given by eqns (9.1) and (10.7) respectively. Equation (9.1) may be expressed:

$$1.79(ArDe)^{0.5} \cdot \frac{\varepsilon_{mb}^{1-n}}{nRe_t(1-\varepsilon_{mb})^{0.5}} = 1. \qquad (13.17)$$

Since both n and Re_t are functions of Ar (eqns (4.5) and (2.17) respectively), eqn (13.17) shows ε_{mb} to be a function solely of Ar and De, or equivalently of Ga and De.

Equation (10.7) may be expressed in dimensionless form:

$$\left[\frac{d_p \Delta \alpha}{u_t}\right] = 0.67(ArDe)^{0.5} \cdot \frac{2(n-1) - \varepsilon_{mb}(2n-1)}{Re_t \varepsilon_{mb}(1-\varepsilon_{mb})^{0.5}}, \qquad (13.18)$$

where the term in brackets on the left-hand side represents the dimensionless amplitude growth rate parameter, which, for the same reason as for ε_{mb}, is also a function solely of Ga and De. Equations (13.17) and (13.18) thus link the *dimensionless* fluidization quality parameters, ε_{mb} and $d_p\Delta a/u_t$, to corresponding scaling parameters Ar and De, or Ga and De: there is complete compatibility of the two approaches. This, of course, has to be so because the scaling laws guarantee the equivalence of all scaled quantities in matched dimensionless systems.

It is only when fluidization quality matching is made on the basis of the *unscaled* growth rate parameter Δa that different equivalent systems can be identified. An example application is reported in some detail in Chapter 16: two systems, matched in terms of ε_{mb} and Δa, are compared by means of two-dimensional numerical simulation. It will be seen that the resulting equivalence occurs in real (process) time rather than in dimensionless time, which would have been the case if the matching had been carried in accord with the scaling relations as described in this chapter.

The long- and short-wave equations

In addition to the practical applications considered so far in this chapter, scaling can represent a powerful analytical tool in its own right; in particular for reducing general relations to simplified forms that are applicable under restricted conditions.

Consider the equation derived in Chapter 8 describing the propagation of small void fraction perturbations in a fluidized bed:

$$\frac{\partial^2 \varepsilon^*}{\partial t^2} - u_D^2 \frac{\partial^2 \varepsilon^*}{\partial z^2} + D\left(\frac{\partial \varepsilon^*}{\partial t} + u_K \frac{\partial \varepsilon^*}{\partial z}\right) = 0. \qquad (13.19)$$

In Chapter 10, expressions for the velocity v and amplitude growth rate a of a perturbation wave satisfying eqn (13.19) were presented as functions of wavelength λ: eqns (10.1) and (10.3) respectively. These showed that long and short waves propagate at the kinematic- and dynamic-wave speeds, u_K and u_D respectively. Scaling, as will now be demonstrated, enables specific equations for long- and short-wave propagation to be obtained from the more general relation of eqn (13.19).

This time we use wavelength λ as the length reference level, so that dimensional quantities in eqn (13.19) may be related to dimensionless ones through:

$$z = \hat{z} \cdot \lambda, \quad t = \hat{t} \cdot \lambda / u_t, \quad u_K = \hat{u}_K \cdot u_t, \quad u_D = \hat{u}_D \cdot u_t. \tag{13.20}$$

Making these substitutions in eqn (13.19) yields the dimensionless relation:

$$\frac{\partial^2 \varepsilon^*}{\partial \hat{t}^2} - \hat{u}_D^2 \frac{\partial^2 \varepsilon^*}{\partial \hat{z}^2} + \left(\frac{D\lambda}{u_t}\right)\left(\frac{\partial \varepsilon^*}{\partial \hat{t}} + \hat{u}_K \frac{\partial \varepsilon^*}{\partial \hat{z}}\right) = 0. \tag{13.21}$$

Long waves

For large $(D\lambda/u_t)$ eqn (13.21) reduces to:

$$\frac{\partial \varepsilon^*}{\partial \hat{t}} + \hat{u}_K \frac{\partial \varepsilon^*}{\partial \hat{z}} = 0, \tag{13.22}$$

which converts to the same dimensional form:

$$\frac{\partial \varepsilon^*}{\partial t} + u_K \frac{\partial \varepsilon^*}{\partial z} = 0. \tag{13.23}$$

This equation describes the propagation of kinematic waves, as may be readily verified from the expression for a travelling wave, eqn (7.18), and by proceeding exactly as illustrated in eqns (7.19)–(7.22); the wave solution to eqn (13.23) travels, without change of amplitude, at the *kinematic-wave* speed: $v = u_K$, $a = 0$.

Short waves

For small $(D\lambda/u_t)$ eqn (13.21) reduces to the equation for dynamic-wave propagation:

$$\frac{\partial^2 \varepsilon^*}{\partial t^2} - u_D^2 \frac{\partial^2 \varepsilon^*}{\partial z^2} = 0. \tag{13.24}$$

The wave solution to eqn (13.24) follows from the same procedure adopted above: it describes in this case the propagation of constant amplitude *dynamic waves*: $v = \pm u_D$, $a = 0$. It should be pointed out, however, that this convergence of eqn (13.21) to eqn (13.24) is short-lived: for the stable case, $u_K^2 < u_D^2$, the disturbances initially propagate at dynamic-wave speeds (second order characteristic velocities) $v = \pm\, u_D$, governed by eqn (13.24); but these damp out exponentially with

$$\exp\left(-\frac{u_K}{u_D}\cdot\frac{u_t}{D\lambda}\cdot t\right),$$

and the main disturbance lags behind and propagates at the kinematic-wave speed u_K. Full details of this, and of wave propagation patterns in general that are governed by eqn (13.19), are given by Whitham (1974).

References

Crowther, M.E. and Whitehead, J.C. (1978). Fluidization of fine powders at elevated pressure. In: *Fluidization* (J.F. Davidson and D.L. Keairns, eds). Cambridge University Press.

Di Felice, R., Rapagnà, S. and Foscolo, P.U. (1992). Dynamic similarity rules: validity check for bubbling and slugging beds. *Powder Technol.*, **71**, 281.

Fitzgerald, T., Bushnell, D., Crane, S. and Shieh, Y.-C. (1984). Testing of cold scaled modelling for fluidized bed combustors. *Powder Technol.*, **38**, 107.

Foscolo, P.U., Germanà, A., Di Felice, R. *et al.* (1989). An experimental study of the expansion characteristics of fluidized beds of fine catalysts under pressure. In: *Fluidization VI* (J.R. Grace, L.W. Shemilt and M.A. Bergougnou, eds). Engineering Foundation.

Foscolo, P.U., Gibilaro, L.G., Di Felice, R. *et al.* (1990). Scaling relations for fluidization: the generalized particle bed model. *Chem. Eng. Sci.*, **45**, 1647.

Foscolo, P.U., Gibilaro, L.G. and Di Felice, R. (1991). Hydrodynamic scaling relationships for fluidization. *Appl. Sci. Res.*, **48**, 315.

Gibilaro, L.G., Hossain, I. and Foscolo, P.U. (1986). Aggregate behaviour of liquid-fluidized beds. *Can. J. Chem. Eng.*, **64**, 931.

Gibilaro, L.G., Di Felice, R., Foscolo, P.U. and Waldram, S.P. (1988). Fluidization quality: a criterion for indeterminate stability. *Chem. Eng. J.*, **37**, 25.

Glicksman, L.R. (1984). Scaling relationships for fluidized beds. *Chem. Eng. Sci.*, **39**, 1373.

Grace, J. (1986). Contacting modes and behaviour of gas-solid and other two-phase suspensions. *Can. J. Chem. Eng.*, **64**, 353.

Grace, J. and Harrison, D. (1970). Design of fluidized beds with internal baffles. *Chem. Proc. Eng.*, **46**, 127.

Jackson, R. (1963). The mechanics of fluidized beds: Part 1: The stability of the state of uniform fluidization. *Trans. Inst. Chem. Eng.*, **41**, 13.

Jacob, K.V. and Weimer, A.W. (1987). High pressure particulate expansion and minimum bubbling of fine carbon powders. *AIChE J.*, **33**, 1698.

Massey, B.S. (1971). *Units, Dimensional Analysis and Physical Similarity.* London: Van Nostrand Reinhold.

Rapagnà, S., Di Felice, R., Foscolo, P.U. and Gibilaro, L.G. (1992). Experimental verification of the scaling rules for fine powder fluidization. *Fluidization VII* (O.E. Potter and D.J. Nicklin, eds). Engineering Foundation.

Whitham, G.B. (1974). *Linear and Non-Linear Waves.* John Wiley & Sons.

14

The jump conditions

Large perturbations in fluidized beds

The analyses reported in Chapters 6 to 11 focused on fluidized beds subjected to *small* perturbations, which could be treated in terms of the *linearized* conservation equations for mass and momentum. This was perfectly valid for homogeneously fluidized systems right up to the *minimum bubbling point*, where perturbations start to grow and homogeneity is destroyed. For bubbling fluidization, on the other hand, linear analysis is inadequate: the bubbles in this case represent *large* perturbations in void fraction, even discontinuities or *shocks*, separating a dense particle phase from the completely void, or nearly completely void, bubble phase: the conditions for linearity are clearly violated (Figure 14.1).

In this chapter we apply the *unlinearized particle bed* model to the study of discontinuities in fluidized beds (Brandani and Foscolo, 1994; Sergeev *et al.*, 1998). The *jump conditions* will be seen to supply remarkably straightforward and elegant means for analysing such occurrences. Earlier studies of the general

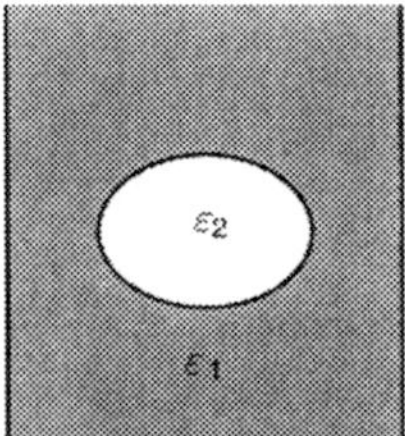

Figure 14.1 Large void fraction perturbations in fluidized beds.

problem have been reported by Buyevich and Gupalo (1970) and Fanucci *et al.* (1981). The jump conditions have also been applied to systems in which the particles are subjected to magnetic, in addition to fluid-dynamic, forces (Brandani and Astarita, 1996; Sergeev and Dobritsyn, 1995). This latter paper, together with that of Harris and Crighton (1994), contains a fairly comprehensive list of references to mathematical publications on the analysis of non-linear wave propagation and the formation of discontinuities in fluidized beds based on systems of equations similar to those employed in this book. The much simpler analysis that now follows will be shown to lead to verifiable predictions of bubbling and slugging behaviour, rationalizing a number of empirically well-known phenomena.

The jump conditions

Consider the one-dimensional situation, depicted in Figure 14.2, of a shockwave propagating upwards, with velocity V, through a fluidized suspension. The void fractions immediately below and above the shock are ε_1 and ε_2 respectively. (The jump condition derivations that now follow are less restricted than appears from Figure 14.2, in that the void fractions directly across the shock, ε_1 and ε_2, need not in general correspond to equilibrium conditions.)

Necessary conditions for the existence of a shock are that mass and momentum are conserved across it. In order to quantify these conditions, we will apply the simplified *particle bed* model formulation introduced in Chapter 8, in which the condition $\rho_p \gg \rho_f$ enabled the particle-phase equations to be treated independently: eqns (8.21) and (8.25). These are reproduced below, eqns (14.1) and (14.2), with the expression for the dynamic-wave velocity, eqn (8.19), inserted in the elasticity term of

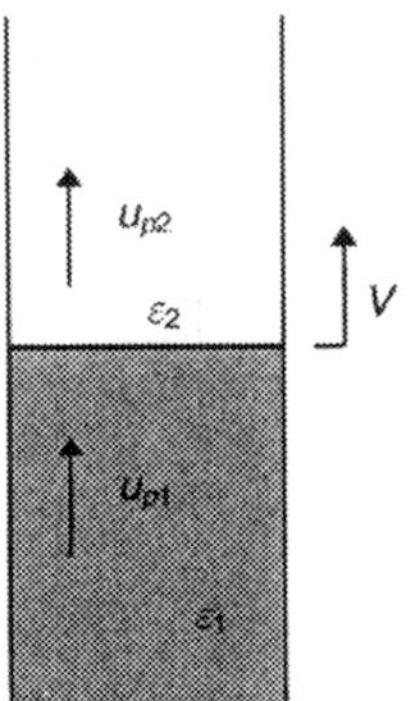

Figure 14.2 A one-dimensional shockwave.

eqn (14.2). The approximation to the full set of defining equations is clearly justified for all cases of gas fluidization, and represents a working approximation for water fluidization of relatively dense particles – as was shown to be the case for the linearized system in the opening section of Chapter 12.

$$\frac{\partial \varepsilon}{\partial t} - \frac{\partial}{\partial z}\left(u_p(1-\varepsilon)\right) = 0, \tag{14.1}$$

$$(1-\varepsilon)\rho_p\left[\frac{\partial u_p}{\partial t} + u_p\frac{\partial u_p}{\partial z}\right] = F(u_p, \varepsilon) + 3.2gd_p(1-\varepsilon)(\rho_p - \rho_f)\frac{\partial \varepsilon}{\partial z}, \tag{14.2}$$

where

$$F(u_p, \varepsilon) = \frac{F_d}{\varepsilon} - (1-\varepsilon)\rho_p g.$$

Derivation of the jump conditions

Integration of eqns (14.1) and (14.2) *across the shock* yields the jump conditions relating the variables on one side to those on the other. However, it is not at all obvious at first sight how these integrations are to be carried out. We now apply a general method, applicable to a wide class of hyperbolic partial differential equations, which was proposed back in the 1920s by the Russian mathematician N.E. Kotchine (1926). The method may be illustrated very cleanly by considering the *steady state* situation of the propagating shock that is described by eqns (14.1) and (14.2) on removal of the time-derivative terms.

We first adopt a co-ordinate system in which the shock is brought to rest, thereby defining new velocity and distance variables:

$$\begin{aligned} \breve{u}_{p1} &= u_{p1} - V, \\ \breve{u}_{p2} &= u_{p2} - V, \\ \breve{z} &= z - \int V\,dt. \end{aligned} \tag{14.3}$$

On this basis, the steady-state mass and momentum equations become:

$$\frac{d}{d\breve{z}}\left(\breve{u}_p(1-\varepsilon)\right) = 0, \tag{14.4}$$

$$\rho_p(1-\varepsilon)\breve{u}_p\frac{d\breve{u}_p}{d\breve{z}} = F(\breve{u}_p,\varepsilon) + 3.2gd_p(1-\varepsilon)(\rho_p-\rho_f)\frac{d\varepsilon}{d\breve{z}}. \tag{14.5}$$

The first jump condition can now be obtained by direct integration of eqn (14.4):

$$\breve{u}_p(1-\varepsilon) = \text{constant}, \quad \breve{u}_{p1}(1-\varepsilon_1) = \breve{u}_{p2}(1-\varepsilon_2); \tag{14.6}$$

the second requires application of the general method as follows.

First write the integral of eqn (14.5) over the distance interval 2δ that includes the shock at its centre:

$$\begin{aligned} &\rho_p\int_{-\delta}^{+\delta}\left((1-\varepsilon)\breve{u}_p\frac{d\breve{u}_p}{d\breve{z}}\right)d\breve{z} \\ &= \int_{-\delta}^{+\delta} F\,d\breve{z} + 3.2gd_p(\rho_p-\rho_f)\int_{-\delta}^{+\delta}\left((1-\varepsilon)\frac{d\varepsilon}{d\breve{z}}\right)d\breve{z}. \end{aligned} \tag{14.7}$$

Next consider the limit as $\delta \to 0$. Terms that do not contain a z-derivative vanish. The z-derivatives of variables that experience a jump approach infinite values, resulting in finite limits for the integrals:

$$\rho_p\int_{\breve{u}_{p1}}^{\breve{u}_{p2}}(1-\varepsilon)\breve{u}_p d\breve{u}_p = 3.2gd_p(\rho_p-\rho_f)\int_{\varepsilon_1}^{\varepsilon_2}(1-\varepsilon)d\varepsilon. \tag{14.8}$$

Note that the product $(1-\varepsilon)\breve{u}_p$ in the left-hand integral of eqn (14.8) is a constant, provided by the continuity jump condition, eqn (14.6). Integration of eqn (14.8) then yields the second jump condition, eqn (14.10).

The jump conditions:

$$\left[(1-\varepsilon)\breve{u}_p\right] = 0; \tag{14.9}$$

$$\rho_p\left\{(1-\varepsilon)\breve{u}_p\right\}_{1,2}\cdot\left[\breve{u}_p\right] - 1.6gd_p(\rho_p-\rho_f)[\varepsilon(2-\varepsilon)] = 0. \tag{14.10}$$

The square brackets in eqns (14.9) and (14.10) denote the jump across the shock of the quantity included within them: $[A] = A_1 - A_2$, and the subscript 1,2 indicates that all quantities within the bracket to which it refers may be evaluated either behind or in front of the shock.

It must be emphasized at this point that the steady-state assumption in the above analysis was introduced solely for the purpose of uncluttering the presentation. The full unsteady-state mass and momentum equations could equally well have been used. They lead to the same jump conditions, eqns (14.9) and (14.10). This is because the omitted time derivative terms do not involve a jump, and therefore vanish when the integration interval is reduced to the infinitesimal limit.

The shock velocity

On writing the continuity jump condition, eqn (14.9), in terms of velocities relative to the bed wall,

$$(1-\varepsilon_1)(u_{p1}-V) = (1-\varepsilon_2)(u_{p2}-V), \tag{14.11}$$

we obtain, on rearrangement:

$$V - u_{p1} = -(1-\varepsilon_2)\cdot\frac{(u_{p1}-u_{p2})}{(\varepsilon_1-\varepsilon_2)}. \tag{14.12}$$

Another expression for $V - u_{p1}$ follows from the momentum jump condition, eqn (14.10):

$$V - u_{p1} = -\frac{3.2gd_p(\rho_p-\rho_f)}{\rho_p}\cdot\frac{(2-\varepsilon_1-\varepsilon_2)}{2(1-\varepsilon_1)}\cdot\frac{(\varepsilon_1-\varepsilon_2)}{(u_{p1}-u_{p2})}. \tag{14.13}$$

Multiplying eqn (14.12) by eqn (14.13), to eliminate the particle velocity jump, and expressing the result in terms of the dynamic-wave velocity before the shock,

$$u_{D1} = \sqrt{3.2gd_p(1-\varepsilon_1)(\rho_p-\rho_f)/\rho_p}, \tag{14.14}$$

yields simple expressions for shock velocity in terms of the void fractions in front and behind:

$$V = u_{p1} \pm u_{D1}\sqrt{\frac{(1-\varepsilon_1)(1-\varepsilon_2)+(1-\varepsilon_2)^2}{2(1-\varepsilon_1)^2}}, \tag{14.15}$$

and, by symmetry:

$$V = u_{p2} \pm u_{D2}\sqrt{\frac{(1-\varepsilon_1)(1-\varepsilon_2)+(1-\varepsilon_1)^2}{2(1-\varepsilon_2)^2}}. \tag{14.16}$$

Equations (14.15) and (14.16) reduce to the familiar form for infinitesimal perturbations, $\varepsilon_1 \to \varepsilon_2 \to \varepsilon$, which travel, relative to the particle phase, at the dynamic-wave velocity u_D:

$$V - u_p = \pm u_D. \tag{14.17}$$

Shock stability

The jump conditions represent only *necessary* conditions for a shock to exist. A further necessary condition is that it must be stable, in the sense that potential disturbances to the shock front are contained. Small such disturbances start to propagate from both sides of the shock at the appropriate dynamic-wave speed. The condition for containment, which prevents this propagation from taking place, is simply that dynamic waves behind the shock travel faster, and those in front travel slower, than the shock itself: in this way all perturbations run towards the shock, thereby preserving its integrity (Ganser and Lightbourne, 1991):

Criteria for shock stability:

$$V - u_{p1} < u_{D1}, \quad V - u_{p2} > u_{D2}. \tag{14.18}$$

We are now in a position to answer some important questions regarding shock propagation. The first concerns its direction, whether *upward* or *downward*; and then, for each of these directions, whether the shock gives rise to an *expansion* or a *compression* of the particle phase over which it passes (Figure 14.3).

It is convenient to write the shock velocity expressions, eqns (14.15) and (14.16), in terms of γ, the ratio of particle concentration in front of the shock to that behind it:

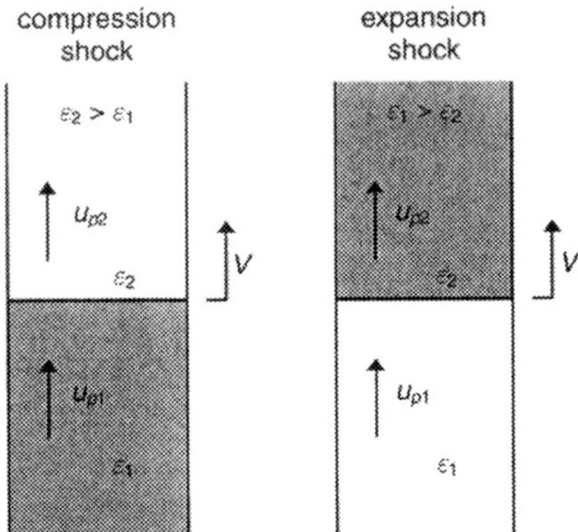

Figure 14.3 Compression and expansion shockwaves.

$$\gamma = \frac{1-\varepsilon_2}{1-\varepsilon_1}, \tag{14.19}$$

$$V - u_{p1} = \pm u_{D1}\sqrt{\frac{\gamma(1+\gamma)}{2}}, \tag{14.20}$$

$$V - u_{p2} = \pm u_{D2}\sqrt{\frac{\gamma+1}{2\gamma^2}}. \tag{14.21}$$

Upwards travelling shocks

The above relations enable shock stability criteria to be expressed very simply. For upwards travelling shocks we apply the positive alternatives on the right of eqns (14.20) and (14.21). The stability criteria, eqn (14.18), then becomes:

$$\gamma(1+\gamma) < 2, \quad (1+\gamma) > 2\gamma^2. \tag{14.22}$$

It is clear that both of these conditions are *always* satisfied for *compression shocks*, $\gamma < 1$, and *never* satisfied for *expansion shocks*, $\gamma > 1$. These conclusions have relevance for slugging fluidization, which will be explored in the following chapter.

Downwards travelling shocks

For this case we apply the negative alternatives in eqns (14.20) and (14.21), giving for the stability criteria:

$$-\gamma(1+\gamma) < 2, \quad -(1+\gamma) > 2\gamma^2. \tag{14.23}$$

The second of these criteria can never be satisfied for any positive value of γ: neither compression nor expansion shocks may propagate downwards in fluidized beds.

The net result of this analysis is quite definite and remarkably simple: of all possible shocks that satisfy the jump conditions, eqns (14.9) and (14.10), all and only *upwards travelling compression shocks* satisfy further necessary conditions for existence.

Compatibility of the jump conditions with the linear stability analysis

Equations (14.12) and (14.13) represent shock velocities relative to the particle phase, evaluated on the basis that mass and momentum respectively are conserved. That these two velocities should be the same may be regarded as a necessary condition for the shock to exist. Equating them yields a statement of this condition, which becomes:

$$\frac{3.2gd_p(\rho_p - \rho_f)}{\rho_p} \cdot \frac{(2 - \varepsilon_1 - \varepsilon_2)}{2} = (1 - \varepsilon_1)(1 - \varepsilon_2)\left(\frac{u_{p1} - u_{p2}}{\varepsilon_1 - \varepsilon_2}\right)^2. \tag{14.24}$$

If we now consider the limiting condition for eqn (14.24) of an infinitesimal jump ($\varepsilon_1 \to \varepsilon_2 \to \varepsilon$, $u_{p1} \to u_{p2} \to u_p$), it reduces to:

$$\frac{3.2gd_p(\rho_p - \rho_f)(1 - \varepsilon)}{\rho_p} = \left(-(1 - \varepsilon) \cdot \frac{du_p}{d\varepsilon}\right)^2. \tag{14.25}$$

This relation is precisely Wallis's criterion for the linear stability limit for homogeneous fluidization (the minimum bubbling point), the left- and right-hand sides comprising the squares of, respectively, the familiar forms for the dynamic-wave speed u_{D}, and the kinematic-wave speed u_{K}:

$$u_{\mathrm{D}}^2 = u_{\mathrm{K}}^2. \tag{14.26}$$

(The general expression for kinematic-wave speed relative to the particle-phase is given by: $u_{\mathrm{K}} = (1 - \varepsilon)d(U_0 - u_p)/d\varepsilon$. In Chapter 5 this relation was derived for the case of constant u_p, eqn (5.9); in eqn (14.25) it corresponds to constant U_0. The familiar explicit form, $u_{\mathrm{K}} = nu_t(1 - \varepsilon)\varepsilon^{n-1}$, emerges on evaluating the derivative term from the empirical Richardson–Zaki law: $U_0 - u_p = u_t\varepsilon^n$.)

In showing how the minimum bubbling point can be determined from an examination of the structure of *heterogeneous* fluidization, as an alternative to the traditional approach based on analysis of the *homogeneously* fluidized state, we have both confirmed the internal consistency of the *particle bed* model formulation, and generalized the Wallis criterion to encompass finite perturbations.

Prediction of void fraction jump magnitudes

We are now well along the way to addressing a number of important basic questions concerning heterogeneous fluidization: in particular, given that we know or can estimate the void fraction in a fluidized dense-phase, what does the theory tell us about the void fraction to expect in an inhomogeneity, or bubble? Empirical observations are quite definite on this matter: for gas-fluidized group B powders, the dense-phase void fraction of about 0.4 gives way to a virtually completely void bubble phase; for the finer, group A powders the bubbles have been reported to contain a few percent of solids, which contribute to the good performance of these systems as chemical reactors (Grace and Sun, 1991); for liquid fluidization, 'parvoid' inhomogeneities containing significant quantities of solids have been widely reported and are discussed in Chapter 12.

Up to now we have been able to analyse aspects of shock behaviour solely on the basis of the jump conditions themselves. These provide two equations in terms of the five variables needed to fully define a shock (ε_1, ε_2, u_{p1}, u_{p2}, V). To proceed further it becomes necessary to specify other relations linking these variables. This can be done with the assumption of dynamic equilibrium on each side of the shock. Expressing this condition in terms of the Richardson–Zaki law,

$$U_0 - u_{p1} = u_t \varepsilon_1^n, \quad U_0 - u_{p2} = u_t \varepsilon_2^n, \tag{14.27}$$

and using these relations to substitute for $u_{p1} - u_{p2}$ in eqn (14.24), we obtain the following expression in terms of the dimensionless Froud and density numbers:

$$\frac{(1 - De)}{Fr} - \frac{0.625(1 - \varepsilon_1)(1 - \varepsilon_2)}{2 - \varepsilon_1 - \varepsilon_2}\left(\frac{\varepsilon_1^n - \varepsilon_2^n}{\varepsilon_1 - \varepsilon_2}\right)^2 = 0. \tag{14.28}$$

Equation (14.28) enables us to answer the question posed at the start of this section. For any chosen system (which specifies Fr, De and n) having dense-phase void fraction ε_1, the 'bubble' void fraction ε_2 may now be obtained by iteration.

Group B systems: verification of the 'two-phase theory' for gas fluidization

Figure 14.4 contains solutions to eqn (14.28) for a typical group B air-fluidized system: it shows bubble void fractions ε_2 for all values of dense phase void fraction ε_1. There is a lot of information in this diagram, but only one solution is of relevance from a strictly practical point of view: the jump at the minimum fluidization condition, $\varepsilon_1 = \varepsilon_{mf} = 0.4$, to a virtually completely void bubble, $\varepsilon_2 \approx 1$. (In fact this value computes to over 0.999.) This result provides a truly theoretical justification for the long established 'two-phase theory' of gas fluidization for moderately sized powders, which postulates a dense particle phase that remains at the minimum bubbling condition for all fluid fluxes in excess of U_{mf}, with the remaining gas forming completely void bubbles (Toomey and Johnstone, 1952)

Other features of the void fraction jump characteristics illustrated in Figure 14.4, although of no direct physical relevance for group B powders, become important for other systems. The void fraction at the minimum bubbling point ε_{mb} occurs in the physically unrealizable region

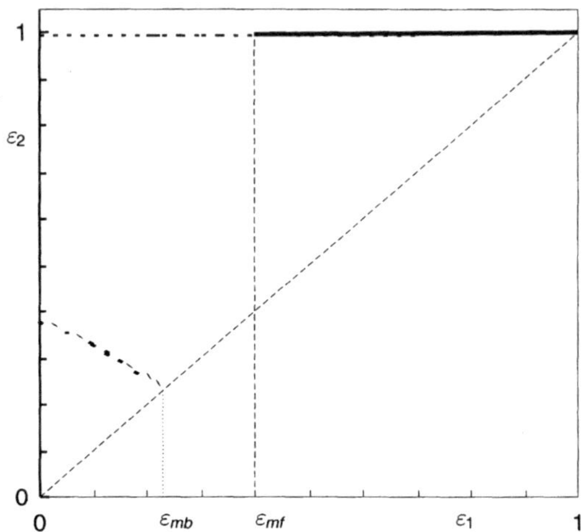

Figure 14.4 Void fraction jumps for ambient air fluidization of a typical Geldart group B powder: $\mu_f = 1.8 \times 10^{-5}\,\mathrm{Ns/m^2}$, $\rho_f = 1.2\,\mathrm{kg/m^3}$, $\rho_p = 1500\,\mathrm{kg/m^3}$, $d_p = 200\,\mu\mathrm{m}$.

to the left of ε_{mf}, where all features are shown as broken lines. To the left of ε_{mb} there are always two solution curves for ε_2: one branching from the diagonal to the left, at ε_{mb} itself; the other remaining at a value very close to unity. This latter curve proceeds rightwards, passing $\varepsilon_1 = \varepsilon_{mf}$, where it comes to represent a physically realizable jump to an almost completely void bubble, and then on to the fully expanded bed limit, $\varepsilon_1 = 1$.

Group A systems

A typical ambient gas fluidized fine-powder system is illustrated in Figure 14.5. This time the minimum bubbling condition is a physical reality, and possible patterns of behaviour are more complex. To the left of ε_{mb} there are again two solutions for the jump to ε_2, but this time the two branches are revealed as a single curve, with a 'nose' just to the right of the ε_2 axis. In the region between ε_{mf} and ε_{mb}, we have at least two physically realizable possibilities: homogeneous expansion, shown as a continuous line along the diagonal, and a jump somewhere within this region to a high void fraction ($\varepsilon_2 = 0.98$) at $\varepsilon_1 = \varepsilon_{mf}$.

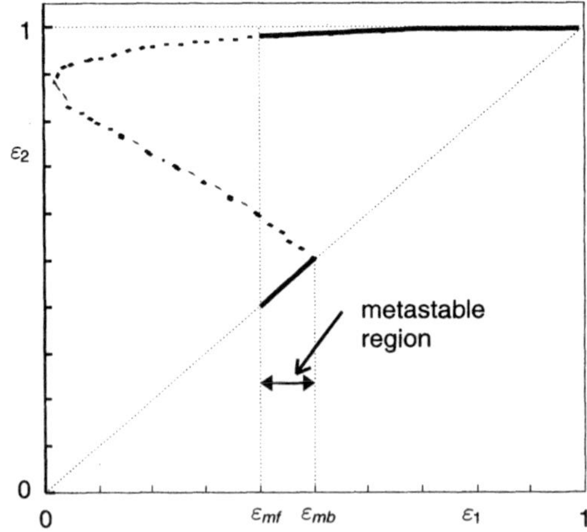

Figure 14.5 Void fraction jumps for ambient air fluidization of a typical Group A powder: $\mu_f = 1.8 \times 10^{-5}$ Ns/m^2, $\rho_f = 1.2$ kg/m^3, $\rho_p = 1000$ kg/m^3, $d_p = 80$ μm.

The metastable fluidized state

The above results provide a theoretical explanation for the experimental observation, referred to at the start of this chapter, of a few per cent of particles (up to 2 per cent for this example) in group A system bubbles; also for the well-documented, *metastable* condition of fluidized beds in the homogeneous expansion region between ε_{mf} and ε_{mb} (Abrahamsen and Geldart, 1980): as discussed in Chapter 9, extreme care must be taken in determining ε_{mb} experimentally, because any imposed disturbance (due to a flow obstruction caused by a thermometer pocket, for example) has the effect of driving the bed prematurely into the bubbling state. This phenomenon can now be clearly explained with reference to Figure 14.5. In the metastable region the homogeneous state is stable to *small* perturbations, as was established by the *linear* analysis of Chapter 8: homogeneous expansion proceeds through this region along the diagonal as depicted. Perturbations that exceed the linear response limits, however, give rise to a jump to the high void fraction depicted by the upper continuous curve, as revealed by the *non-linear* analysis described in this chapter.

Yet another theoretical possibility for the metastable region is a jump to the relatively low void fraction 'bubble' represented by the broken line which branches to the left from the diagonal at the minimum bubbling point shown in Figure 14.5. It seems unlikely, however, that this condition would ever be observed in practice, as the upper solution entails a lower potential energy condition than does the lower one, and therefore represents the more stable outcome.

Bed collapse at the minimum bubbling point

The collapse, or sudden height reduction, of a fine powder fluidized bed on attaining the minimum bubbling point is a widely reported experimental phenomenon. It may be attributed entirely to the metastable condition described above. Consider a bed carefully expanded homogeneously across the metastable region by progressively increasing the fluid flux U_0, leading to a progressive increase in bed height H. On reaching the minimum bubbling point, bubbles start to appear, which represent *large* perturbations to which the expanded homogeneous state at ε_{mb} is no longer stable: the bed collapses to a lower void fraction by the evolution of more bubbles, resulting in yet further instability, and so on (Figure 14.6). For the group A example considered above, dense phase contraction right back to ε_{mf} is a distinct possibility. Subsequent bed expansion, for fluid fluxes above U_{mb}, is simply a result of increasing bubble holdup.

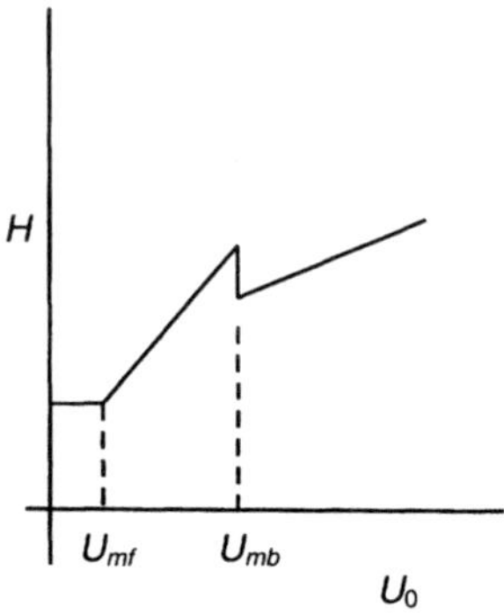

Figure 14.6 Bed collapse at the minimum bubbling point.

The effect of high fluid pressure

An observed effect of increasing the fluid pressure is to reduce bed contraction at the minimum bubbling point. Specific experimental investigations into this phenomenon are reported below. Once again, the void fraction jump characteristics furnish theoretical predictions of this observed behaviour. Figure 14.7 represents the group A system considered above, except that the gas density has been increased from $1.2\,kg/m^3$ to $100\,kg/m^3$ to correspond to a high gas pressure. The effect

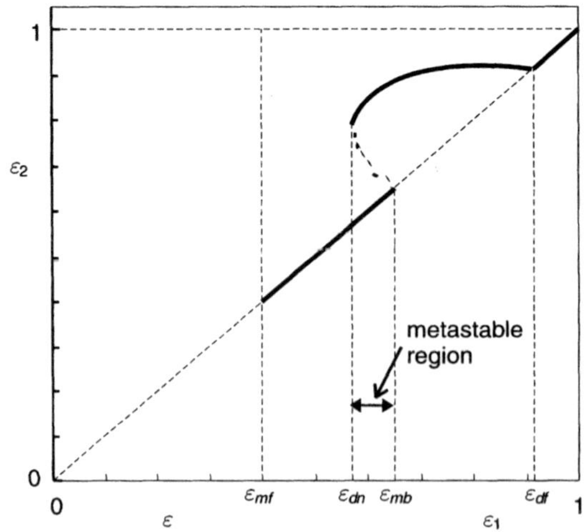

Figure 14.7 Void fraction jumps for high fluid-pressure fluidization of a fine powder: $\mu_f = 1.8 \times 10^{-5}\,Ns/m^2$, $\rho_f = 100\,kg/m^3$, $\rho_p = 1000\,kg/m^3$, $d_p = 80\,\mu m$.

has been to increase ε_{mb}, and at the same time to reduce the metastable region to a small segment close to ε_{mb}.

The effect of high fluid pressure is therefore to reduce the extent of possible bed collapse, as is observed in practice. The region from ε_{mf} up to ε_{dn}, the void fraction corresponding to the tip of the 'nose' in Figure 14.7, which marks the start of the metastable region, is unequivocally stable to both infinitesimal and finite perturbations, signifying homogeneous fluidization as the only possibility. A further effect of pressure is to enlarge the *dilute fluidization* region (the second region of stable homogeneous expansion beyond ε_{df}).

Experimental determination of dense-phase void fraction

From the foregoing discussion it would appear that for fluidized beds that exhibit an initial region of homogeneous expansion, the dense-phase void fraction in the subsequent bubbling regime should lie somewhere between ε_{dn} and ε_{mb}, perhaps moving closer to the former limit as the fluid flux, and hence the perturbation intensity, is progressively increased beyond the minimum bubbling point. This represents a model prediction that is readily amenable to experimental examination.

Bed collapse experiments

The dense-phase void fraction ε_d in the bubbling regime may be determined by suddenly shutting off the fluid flux and recording the bed surface height H as it falls with time. This conceptually simple experiment furnishes much information on the bubbling state (Rietema, 1967).

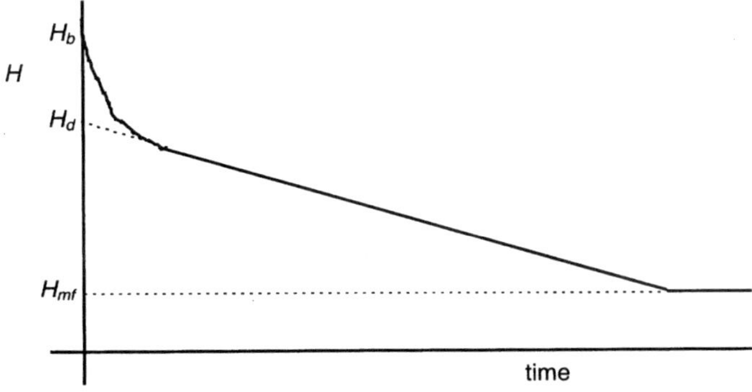

Figure 14.8 Bed collapse experiments.

The essential interpretation of the experimental data may be readily appreciated from the somewhat idealized representation of Figure 14.8: in addition to the dense-phase void fraction ε_d, bubble holdup and interstitial gas flow rates may be estimated from this response.

The initial sharp fall in bed height relates to the escape of bubbles immediately after the gas flux has been cut off. Thereafter, the dense phase collapses linearly with time in the manner of a homogeneous bed as described in Chapter 5. Extrapolating this linear segment back to the H axis yields the height H_d of a notional homogeneous bed from which its void fraction, the required ε_d, follows from a knowledge of the total volume V_p of particles per unit area of bed cross-section: $V_p = H_d(1 - \varepsilon_d)$.

Dense-phase void fractions reported for ambient gas fluidization

Bed collapse experiments were employed, along with other tests, in a comprehensive study of fine powder fluidization under ambient conditions (Foscolo *et al.*, 1987). Four fine powder beds were each fluidized by three gases: air, argon and carbon dioxide. The system properties are given in Table 14.1.

In addition to the ε_d determinations reported below, interstitial gas velocities (from the gradient of the linear portion of the bed surface response) and bubble fractions $(1 - H_d/H_b)$ were obtained from the bed collapse experiments.

For each of the 12 systems studied, the dense-phase void fraction ε_d was found to fall sharply from its value at the minimum bubbling point ε_{mb}.

Table 14.1 Gas and particle properties for bed collapse experiments

Gas properties			*Particle properties*		
	ρ_f (kg/m^3)	μ_f (Ns/m^2 × 10^5)		ρ_p (kg/m^3)	d_p (μm)
Air	1.09	1.8	Circles	873	56
Argon	1.48	2.2	Squares	1054	62
Carbon dioxide	1.64	1.4	Triangles	1500	61
			Diamonds	1650	69

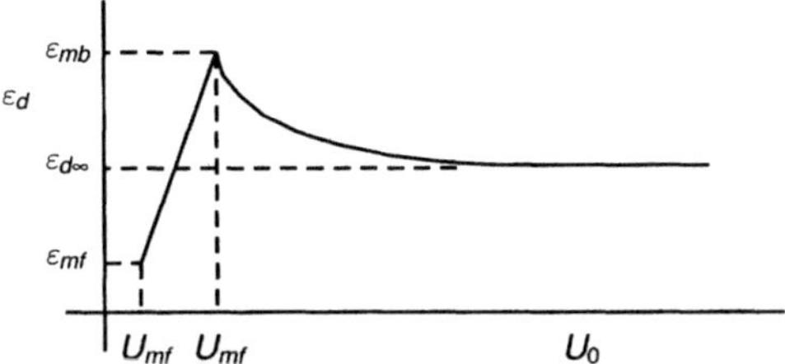

Figure 14.9 Dense-phase void fraction as a function of fluid flux.

Further increases in fluid flux led to progressive reductions in ε_d, which approached a constant value $\varepsilon_{d\infty}$ at gas fluxes in excess of approximately $4U_{mb}$. For all but one of the systems, this limiting void fraction was greater than that at minimum fluidization: $\varepsilon_{mb} > \varepsilon_{d\infty} > \varepsilon_{mf}$. This relation is illustrated in Figure 14.9.

The experimentally determined values for $\varepsilon_{d\infty}$ are shown as points in Figure 14.10. The vertical lines through each point represent the range of possible $\varepsilon_{d\infty}$ values predicted by the *particle bed* model, from the bottom limit of ε_{dn} (corresponding to the 'nose' of the void fraction jump diagram) or ε_{mf}, whichever is the higher, to the upper limit of ε_{mb}. The only experimental point to fall outside the predicted range is the highest value reported, which also corresponds to the highest values for both the upper and lower predicted bounds.

Experiments at elevated fluid pressure

A similar series of tests were performed on four fine powder systems fluidized at pressures ranging from ambient to 30 bar (Foscolo *et al.*, 1989).

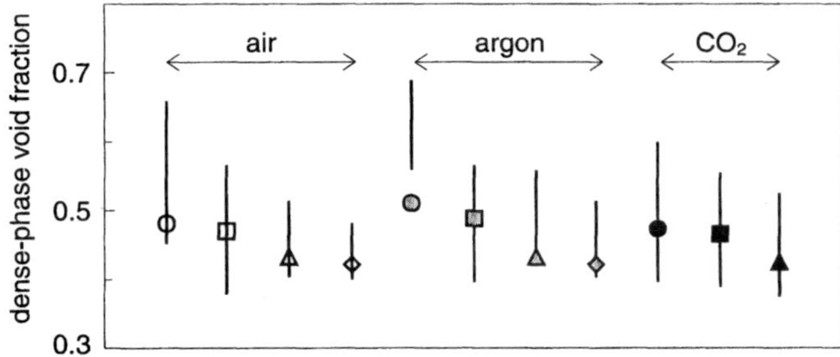

Figure 14.10 Experimentally determined values of $\varepsilon_{d\infty}$ (points) and predicted bounds (vertical lines).

The apparatus described in Chapter 13, which allows for visual observation of a transparent bed through narrow vertical windows in the containing pressure vessel, was used for this purpose. The effect of increasing fluid pressure followed the predicted trend, giving rise to an increase in ε_{mb} of some 10 per cent over the range examined; a correspondingly greater increase in dense-phase void fraction was also observed, again in agreement with model predictions of a decreasing region of metastable behaviour.

The experiments at much higher pressure (up to 120 bar) performed by Jacob and Weimer have already been referred to in Chapter 9 (Jacob and Weimer, 1987). These were also in full agreement with the *particle bed* model. In addition to following ε_{mb} predictions as shown in Figure 9.5, the metastable region was found to reduce progressively with increasing fluid pressure as predicted, eventually vanishing completely to result in zero bed contraction at the minimum bubbling point.

Liquid fluidization

Figure 14.11 shows the void fraction jump characteristics for two examples of ambient water fluidization. The copper powder system on the left is virtually identical to the high-pressure gas-fluidization example considered above. This confirms the equivalence reported for these

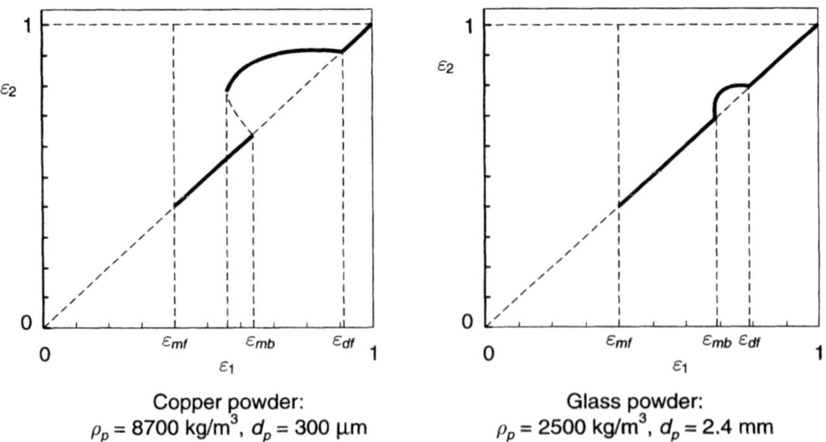

Figure 14.11 Ambient water fluidization of copper and glass powders: $\mu_f = 0.001\ \mathrm{Ns/m^2}$, $\rho_f = 1000\ \mathrm{kg/m^3}$.

systems in Chapter 10 on the basis of the linear, fluidization quality criteria.

The glass powder example on the right represents a system exhibiting a very limited region of instability, with void fraction jumps to values only some 0.1 higher than those of the dense phase. Once again, this is very much in accord with experimental observations of 'parvoids' in such systems; the situation is complicated, however, as a result of low perturbation amplitude decay rates in the homogeneous regime, giving rise to persistent inhomogeneities, discussed in Chapter 12, throughout the entire operating range.

The effect of the jump in fluid pressure

The analysis presented in this chapter has been based on the assumption that particle density is substantially greater than fluid density. This was shown in Chapters 11 and 12 to be valid for all cases of gas fluidization, even under very high-pressure conditions; only liquid-fluidized beds of low-density particles exhibited differences of any significance in the single- and the two-phase treatments. This justification, however, relates to the linear analysis of *small* perturbations. It says little concerning the effect of jumps in fluid pressure across the very considerable discontinuities uncovered in the work described in this chapter. The procedure adopted in Chapter 11, of eliminating fluid-pressure terms by combining the particle- and fluid-phase momentum equations, cannot be utilized here as it involves non-linear manipulations, which are not permitted in the analysis of discontinuous functions.

This problem has been resolved in a satisfactory manner, which is described in detail by Brandani *et al.* (1996). Briefly, it involves evaluating the jump conditions from both the particle- and fluid-phase equations in the manner described above – except that the jumps for terms containing the fluid-pressure gradient, which it is shown may be expressed solely as a function of void fraction, are obtained by numerical integration. The conclusions arising from this two-phase, *non-linear* analysis turn out to be very much in line with those of the two-phase, *linear* analysis described in Chapter 11: for all cases of gas fluidization there are no significant differences in the results of the single- and two-phase treatments; for liquid fluidization differences are generally small for particle densities down to about that of glass (2500 kg/m^3), thereafter increasing with decreasing particle density and increasing particle size.

References

Abrahamsen, A.R. and Geldart, D. (1980). Behaviour of gas-fluidized beds of fine powders. Part 1. Homogeneous expansion. *Powder Technol.*, **26**, 35.

Brandani, S. and Astarita, G. (1996). Analysis of discontinuities in magnetised-bubbling fluidized beds. *Chem. Eng. Sci.*, **51**, 4631.

Brandani, S and Foscolo, P.U. (1994). Analysis of discontinuities arising from the one-dimensional equations of change for fluidization. *Chem. Eng. Sci.*, **49**, 611.

Brandani, S., Rapagnà, S., Foscolo, P.U. and Gibilaro, L.G. (1996). Jump conditions for one-dimensional two-phase shock waves in fluidized beds: the effect of the jump in fluid pressure. *Chem. Eng. Sci.*, **51**, 4639.

Buyevich, Y.A. and Gupalo, Y.P. (1970). Discontinuity surfaces in disperse systems. *Appl. Math. Mech.*, **34**, 722.

Fanucci, J.B., Ness, N. and Yen, R.-H. (1981). Structure of shock waves in gas-particulate fluidized beds. *Phys. Fluids*, **24**, 1944.

Foscolo, P.U., Di Felice, R. and Gibilaro, L.G. (1987). An experimental study of the expansion characteristics of gas fluidized beds of fine catalysts. *Chem. Eng. Prog.*, **22**, 69.

Foscolo, P.U., Germanà, A., Di Felice, R. *et al.* (1989). An experimental study of the expansion characteristics of fluidized beds of fine catalysts under pressure. In: *Fluidization VI* (J.R. Grace, L.W. Shemilt and M.A. Bergougnon, eds), Engineering Foundation.

Ganser, G.H. and Lightbourne, J.H. (1991). Oscillatory travelling waves in a hyperbolic model of a fluidized bed. *Chem. Eng. Sci.*, **46**, 1339.

Grace, J.R. and Sun, G. (1991). Influence of particle size distribution on the performance of fluidized bed reactors. *Can. J. Chem. Eng.*, **69**, 1126.

Harris, S.E. and Crighton, D.G. (1994). Solitons, solitary waves, and voidage disturbances in gas-fluidized beds. *J. Fluid Mech.*, **266**, 243.

Jacob, K.V. and Weimer, A.W. (1987). High-pressure particulate expansion and minimum bubbling of fine carbon powders. *AIChE J.*, **33**, 1698.

Kotchine, N.E. (1926). Sur la thèorie des ondes de choc dans un fluide. *Circ. Mat. Palermo*, **50**, 305.

Rietema, K. (1967). Application of mechanical stress theory to fluidization. *Proc. Int. Symp. Fluidisation, Eindhoven*, p. 154. Netherlands University Press.

Sergeev, Y.A. and Dobritsyn, D.A. (1995). Linear, non-linear small amplitude, steady and shock waves in magnetically stabilized liquid–solid and gas–solid fluidized beds. *Int. J. Multiphase Flow*, **21**, 75.

Sergeev, Y.A., Gibilaro, L.G., Foscolo, P.U. and Brandani, S. (1998). The speed, direction and stability of concentration shocks in a fluidized bed. *Chem. Eng. Sci.*, **53**, 1233.

Toomey, R.D. and Johnstone, H.F. (1952). Gaseous fluidization of solid particles. *Chem. Eng. Prog.*, **48**, 220.

15

Slugging fluidization

The formation of fluid and solid slugs

Up to now the physical dimensions of the vessel containing the fluidized particles have played little or no part in the discussions. For homogeneous fluidization the implicit assumption that the diameter of the bed is much greater than that of a fluidized particle requires little justification for all cases of practical interest. For bubbling fluidization, on the other hand, bubble dimensions can grow to approach that of the bed diameter. In this chapter, we consider the situation in which this limit is reached, and would, but for the presence of the restraining boundary, be overcome: bubbles are then replaced by upwards propagating *fluid slugs*, interspersed with *solid slugs* of the dense phase, both of which extend across the entire bed cross-section. Chaotic bubbling then gives way to more regular oscillatory behaviour, which is characteristic of the *slugging regime*.

Different types of slugging behaviour have been reported and subjected to empirical study, in particular for the case of gas fluidization.

The arguments and analyses presented below apply predominantly to gas systems having particle/bed diameter ratios in excess of about 1/200, which give rise to the formation of *square-nosed* fluid slugs. The boundaries separating adjacent fluid and solid slugs in such systems are essentially horizontal, leading to an overall behaviour pattern that is essentially one-dimensional, amenable to analysis in terms of the one-dimensional equations of change. Lower particle/bed ratios are associated with the formation of *round-nosed, axisymmetric* fluid slugs, around which particles from the dense phase above flow downwards, predominately at the bed wall. High fluid velocities lead to progressively less ordered behaviour: *asymmetric* fluid slugs and *wall slugs*, the latter involving irregular elongated bubbles which travel up the bed wall, effectively signalling an end to truly slugging behaviour and the approach to the turbulent fluidization regime.

The transition from the bubbling to the slugging regime is accompanied by a marked change in fluidization quality, not only with respect to the onset of periodic behaviour. Bubbles provide short cuts for the fluid, enabling it to bypass sections of dense phase in its passage through the bed. In a slugging bed this is no longer the case: short cuts are no longer available, with the result that the fluid residence times are less dispersed – providing some advantage for a reaction environment (Grace and Harrison, 1970).

Relatively shallow beds give rise to a single fluid and solid slug. The periodic nature of slugging in general can be readily appreciated by considering first the somewhat idealized description of single, square-nosed slugging illustrated in Figure 15.1.

At the start of the cycle, the bed consists solely of uniform dense phase. A bubble formed near the distributor grows rapidly to become a fluid slug, occupying the entire tube cross-section as shown in the second

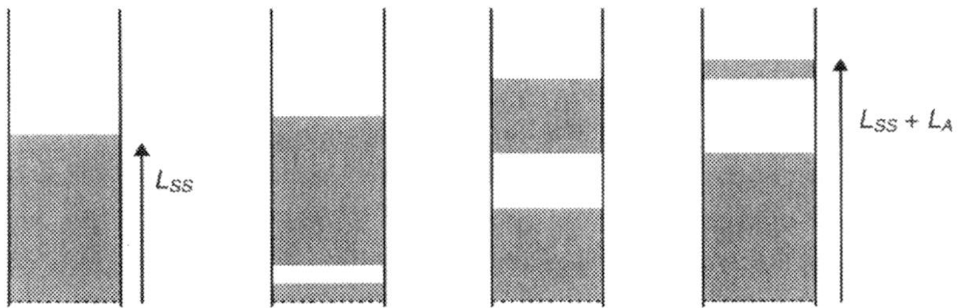

Figure 15.1 Idealized cycle for a single, square-nosed slug.

figure. In the meantime, the solid slug above has been driven, piston-like, some way up the tube. The next figure shows both the fluid slug and the solid slug to have progressed upwards, the former having grown in length and the latter having shrunk, losing particles from its bottom interface to the stagnant dense phase zone below, which remains in contact with the distributor. The final figure shows the fluid slug, at near maximum length, approaching the bed surface, at near maximum elevation – thereafter to fall rapidly to its minimum, starting condition, thereby completing the cycle, which then repeats indefinitely. Deeper beds behave similarly, except that a number of slugs are formed, distributed along the tube length; the top solid slug discharges particles to the one immediately below, which in turn discharges to the next one, and so on down.

The above description is in broad agreement with qualitative observations, and provides a basic structure for analysis. However, it poses more questions than it answers, among which may be included:

- What is it that determines the velocity and initial length of a solid slug?
- What is it that determines the velocity of the gas slug relative to the solid slug?
- Why is it that the bottom segment of dense phase remains in contact with the distributor throughout most of the slugging cycle?
- Why is it that the unrecoverable fluid pressure loss in slugging beds is greater than is observed for other fluidization regimes?
- What role, if any, do particle–particle and particle–wall frictional interactions play in slugging behaviour?

These are some of the questions addressed in the following sections.

An idealized fluid-dynamic description of slugging behaviour

We have good reason to believe that mechanisms other than fluid-dynamic ones play a part in determining the characteristics of slugging fluidization. In Chapter 13 it was reported that, in contrast to the normal situation for homogeneous and bubbling fluidization, the scaling relations fail to deliver conditions for similarity in the slugging regime. Direct particle–particle and particle–wall frictional interactions have been thought to be responsible for this failure, a hypothesis for which we provide convincing support in the following section. Fluid-dynamic mechanisms, on the other hand, provide a theoretical structure, which is able to furnish clear explanations for a number of important aspects of slugging behaviour.

The velocity of a solid slug

For bubbling fluidization of Geldart group B particles, the 'two-phase theory' (Toomey and Johnstone, 1952) postulates a dense phase that remains at the minimum fluidization condition, with excess gas flowing as void bubbles. It was shown in the previous chapter that this outcome is precisely that predicted by the jump-condition analysis for these systems. In slugging operation, the minimum fluidization condition of the dense phase corresponds to the fluid within the solid slug, at void fraction ε_{mf}, travelling at a velocity relative to the particles of U_{mf}/ε_{mf}. Thus we may write: $u_f - u_p = U_{mf}/\varepsilon_{mf}$. We saw in Chapter 7 that for incompressible phases (a reasonable approximation, even for a gas under normal conditions of operation), particle and fluid velocities, u_p and u_f, are linked to the feed flux U_0 by eqn (7.4), from which we obtain: $u_f - u_p = (U_0 - u_p)/\varepsilon_{mf}$. These two relations for the relative fluid–particle velocity yield the simple expression for the velocity u_p of the solid slug:

$$u_p = U_0 - U_{mf}. \tag{15.1}$$

There is abundant evidence that solid slugs generally travel at, or very close to, the value predicted by eqn (15.1).

Potential energy of a solid slug

Consider the hypothetical situation of a single solid slug of length L_{SS} travelling intact up a tube. Forget for the moment the erosion that takes place due to particle shedding from its base. The solid slug is in the state of incipient fluidization, and therefore experiencing a rate of energy dissipation δE_d exactly equal to that of a normal, incipiently fluidized bed:

$$\delta E_d = L_{SS}(1 - \varepsilon_{mf})(\rho_p - \rho_f)gU_{mf}. \tag{15.2}$$

However, the solid slug is also gaining potential energy at a rate δE_{PE} as it rises, at velocity u_p, through the containing tube:

$$\delta E_{PE} = L_{SS}(1 - \varepsilon_{mf})(\rho_p - \rho_f)gu_p. \tag{15.3}$$

These two energy requirements are provided by the fluid, representing a total energy loss in the fluid of $U_0\Delta P$. Equating this quantity to the sum of terms given by eqns (15.2) and (15.3), and applying eqn (15.1), yields the unrecoverable pressure loss in the fluid ΔP:

$$\Delta P = L_{SS}(1 - \varepsilon_{mf})(\rho_p - \rho_f)g. \qquad (15.4)$$

This pressure loss is precisely that which occurs in non-slugging beds, suggesting (quite wrongly, as will shortly be demonstrated) that the potential energy requirement for slugging has no part to play in the phenomenon of progressively increasing pressure loss with fluid velocity, which has long been known to occur under slugging conditions (Baker and Geldart, 1978). It will be seen that this singular feature of slugging behaviour relates critically, albeit indirectly, to the potential energy created during the course of a slugging cycle. Before that, however, a more fundamental property will be examined, which together with eqn (15.1) effectively quantifies the idealized cycle depicted in Figure 15.1: the velocity of a fluid slug.

The velocity of a fluid slug

Fluid slugs travel up the bed faster than solid slugs, progressively eating away the rear of the solid slug as depicted in Figure 15.1. Particles rain down through the fluid slug to end up on the static zone, which remains in contact with the distributor as illustrated; or else, for the case of multi-slug systems, on the solid slug below.

A rising solid slug, in the reference frame that moves with it, may be regarded as an incipiently fluidized bed with a free lower boundary (Gibilaro *et al.*, 1998). The 'bed' lacks a distributor; it is thus convenient to refer to it simply as a *suspended fluidized bed*. Very simple qualitative arguments may be used to demonstrate that the solid slug (or suspended fluidized bed) cannot posses a sharp lower boundary. Imagine such bed having, at an initial moment, a lower boundary consisting of a horizontal, uniformly distributed layer of particles. Consider now the drag force exerted on a particle in this bottom layer by the flowing fluid. It is clear that this force will be smaller than that on a particle situated deeper in the bed because a boundary particle lacks particles below it, so that the fluid velocity, and hence the drag force, experienced at the bottom surface of the boundary particle will be smaller. Another way of expressing this is to say that the total drag on a particle can be decomposed into the sum of two terms; one a function of void fraction at the particle's horizontal centre plane, the other proportional to the *void fraction gradient* across the particle. This second term, which under unsteady-state conditions can affect particles anywhere in the bed, is clearly felt at its strongest by a boundary particle. The effect on the boundary particle is that it immediately starts to move downwards relative to the suspended bed. This displacement rapidly affects particles in the next layer, which respond likewise, so that the particle shedding process progresses continuously upwards. This description is similar to the 'interface stability' argument used in Chapter 5 in relation to the transient response of an expanding, homogeneously fluidized bed.

This 'microscale' consideration shows that the solid slug, or suspended fluidized bed, must be followed by a trail of more dilute particle suspension, starting at void fraction $\varepsilon = \varepsilon_{mf}$ at the lower boundary and followed by progressively increasing void fraction with increasing distance from this boundary.

The interface AA in Figure 15.2 may be regarded as separating the solid slug, all at void fraction ε_{mf}, from the dilute trail. As a result of the steep void fraction gradient experienced by boundary particles, the region immediately below AA, in which the void fraction increases rapidly (the region bounded by AA and BB), is narrow compared to the initial length of the solid slug, and the shed particles very soon reach a velocity close to the unhindered, terminal velocity for free fall under gravity – at correspondingly high void fractions. This is convenient from an experimental point of view, as it means that the lower boundary of the solid slug can be identified visually without ambiguity. The interface AA, in addition to

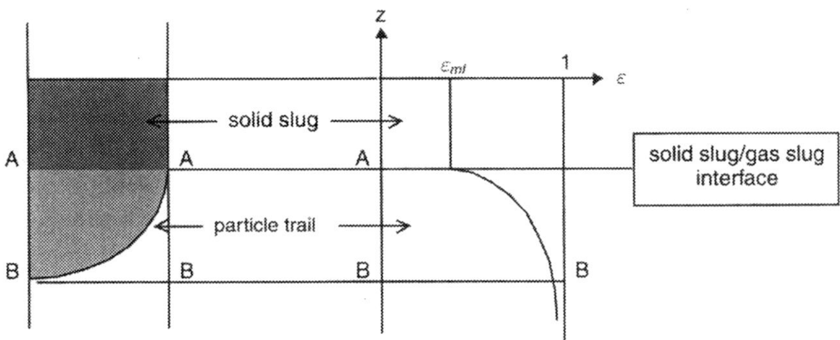

Figure 15.2 The solid slug/fluid slug interface.

representing the lower boundary of the solid slug, may also be regarded as the upper boundary of the fluid slug. The presence of solid particles in the fluid slug is a well-known phenomenon, a result of the particle-shedding process.

The weak shock solution for the lower boundary of the solid slug

The structure of the lower boundary of the solid slug (or suspended fluidized bed) has been discussed above using qualitative, mechanistic reasoning. We can now quantify the conclusions. Should a sharp boundary exist below the solid slug, it could represent an *expansion shockwave* of the particle phase. Although, in contrast to the situation for gas dynamics, there are no thermodynamic equations or constitutive expressions involved that could be used to demonstrate the impossibility of a particle-phase expansion shock, the mathematical form of the one-phase particle bed model, eqns (8.21) and (8.26), is similar to the basic equations of gas dynamics. In the previous chapter it was demonstrated quite unequivocally, on the basis of the *jump conditions* derived from these equations, that an expansion shockwave of the particle phase does not satisfy necessary conditions for its existence – eqns (14.22) and (14.23). This means that the particle concentration across the solid slug lower boundary must be *continuous*, so that the solid slug is followed by a trail in which the particle concentration decreases monotonically – as anticipated above.

In order to analyse the propagation of this lower boundary, the model equations (8.21) and (8.26) must be applied over the entire flow region – that is to say, both within the solid slug itself and in the trail. The lower boundary can then be considered as a *weak discontinuity*, across which the flow variables change continuously (while their first derivatives may, although will not necessarily, suffer a discontinuity). In the reference frame that moves with the solid slug, the propagation velocity of a weak discontinuity coincides with an upward characteristic velocity (Jeffrey, 1976): relative to the solid slug, interface AA travels at the dynamic-wave velocity u_D, and the fluid slug velocity u_{FS} becomes fully predictable:

$$\begin{aligned} u_{FS} &= u_p + u_D = U_0 - U_{mf} + u_D \\ &= U_0 - U_{mf} + \sqrt{3.2gd_p(1-\varepsilon_{mf})(\rho_p-\rho_f)/\rho_p}. \end{aligned} \tag{15.5}$$

For gas-fluidized beds ($\rho_p \gg \rho_f$, $\varepsilon_{mf} \approx 0.4$), eqn (15.5) becomes:

$$u_{FS} = U_0 - U_{mf} + 1.4\sqrt{gd_p}, \quad \textit{Square-nosed gas slugs} \tag{15.6}$$

This theoretical expression for the velocity of square-nosed gas slugs has been shown to be in reasonable agreement with available experimental measurements (Gibilaro *et al.*, 1998). It is of similar form to published correlations (Hovmand and Davidson, 1971) for round-nosed axisymmetic and asymmetric gas slugs, both of which travel much faster relative to the solid slug:

$$u_{FS} = U_0 - U_{mf} + 0.35\sqrt{gD}, \quad \textit{Axisymmetric gas slugs} \tag{15.7}$$

$$u_{FS} = U_0 - U_{mf} + 0.35\sqrt{2gD}. \quad \textit{Asymmetric gas slugs} \tag{15.8}$$

Fluid pressure loss in slugging beds

The rate of energy dissipation, per unit area of bed cross-section, in a gas flowing through a system that gives rise to a total unrecoverable pressure loss ΔP is, by definition, $U_0 \Delta P$. For a fluidized bed, ΔP remains constant (equal to ΔP_B) regardless of the fluid flux U_0. For a slugging bed, however, the progressive increase in ΔP with increasing U_0 is a well-documented phenomenon. Explanations have been suggested in terms of particle–particle and particle wall interactions, and the energy required to accelerate a solid slug to its terminal velocity (Baker and Geldart, 1978). In the experiments reported below we find that the solid frictional effects, which markedly influence other key slugging characteristics, do not contribute to any significant extent to this phenomenon: significant changes in wall and particle roughness make no difference to the observed pressure loss.

We now consider two fluid-dynamic mechanisms for excess fluid pressure loss: kinetic and potential energy requirements of the solid slugs. The arguments will be presented with reference to the simple idealized account of *single square-nosed slug formation and propagation* represented in Figure 15.1.

Kinetic energy requirements of the solid slugs

The kinetic energy required to accelerate a solid slug of length L_{SS} and unit cross-sectional area to its terminal speed u_p at the start of a slugging cycle may be readily quantified for gas-fluidized systems in which 'added mass' effects can be safely discounted: $L_{SS}(1-\varepsilon_{mf})\rho_p u_p^2/2$. At a slugging frequency f, this translates into an unrecoverable pressure loss ΔP_{KE} in the fluid:

$$\Delta P_{KE} = L_{SS}(1-\varepsilon_{mf})\rho_p u_p^2 f/2U_0. \tag{15.9}$$

We shall see later that the excess pressure drop observed in slugging beds can be greatly in excess of that evaluated by means of eqn (15.9).

Potential energy requirements of the solid slugs

Referring to Figure 15.1, we see that over one slugging cycle the solid slug travels a distance equal to the bed surface displacement L_A. Its length decreases progressively over this period, from L_{SS} to zero, so that its potential energy requirement amounts to $L_{SS}(1-\varepsilon_{mf})(\rho_p-\rho_f)gL_A/2$, which translates into an unrecoverable pressure loss ΔP_{PE}:

$$\Delta P_{PE} = L_{SS}(1-\varepsilon_{mf})(\rho_p-\rho_f)gL_A f/2U_0. \tag{15.10}$$

There are a number of things that can be said about this expression. First of all, it usually represents a far larger contribution to the total excess pressure loss than does the corresponding kinetic energy term, eqn (15.9), as can be seen from the ratio:

$$\frac{\Delta P_{PE}}{\Delta P_{KE}} \approx \frac{gL_A}{u_p^2}, \tag{15.11}$$

which is generally large – typically greater than 20 in the experiments we report later in this chapter.

The second observation concerning eqn (15.10) is that a somewhat conservative prediction of this seemingly dominant contribution to excess pressure loss is readily available. In terms of the idealized situation depicted above, the product $L_A f$ is simply the solid slug velocity u_p, equal to $U_0 - U_{mf}$ for most of the slugging cycle. So that:

$$\Delta P_{PE} \approx L_{SS}(1-\varepsilon_{mf})(\rho_p-\rho_f)g \cdot \frac{U_0 - U_{mf}}{2U_0}. \tag{15.12}$$

The first part of this relation for ΔP_{PE} is simply the usual pressure loss for fluidization ΔP_B, reflecting simply suspension of the particles. Hence we may express the excess pressure loss due to potential energy creation in the following compact form:

$$\Delta P_{PE} \approx \frac{U_0 - U_{mf}}{2U_0} \cdot \Delta P_B. \qquad (15.13)$$

This notably simple expression represents a conservative estimate in that it assumes the solid slug velocity to remain constant, taking no account of the acceleration period at the start of the cycle, where the kinetic energy requirements are provided, nor to the inevitable readjustments that occur at the end. Nevertheless, in the experimental programme described below it was found to deliver quite reasonable estimates of the observed excess pressure loss for *square-nosed* slugging systems. For round-nosed systems, on the other hand, eqn (15.13) was found to be more conservative, providing estimates typically double those observed in practice.

Potential energy dissipation

A final observation regarding eqn (15.10) concerns the mechanism linking the potential energy created by rising solid slugs to dissipation in the fluid. This is not immediately obvious, as was the case for the kinetic energy requirement, which simply involved the fluid doing work on a solid slug to accelerate it to its terminal velocity u_p. We saw above, in the arguments leading up to eqn (15.4), that potential energy creation alone does not result in excess pressure loss in the fluid. In slugging operations, however, all the created potential energy is subsequently dissipated. The mechanism whereby this occurs appears to be an indirect one, associated with the particle detachment process at the rear of the solid slug. As the particles rain down through the fluid slug, potential energy converts to kinetic energy, together with some fluid particle frictional dissipation; when they strike the growing region at the base of the bed, particle momentum converts to a downward-acting force on this region, holding it down and thereby enabling it to sustain a gas flux larger than U_{mf} without expansion or upward propagation; this results in an increased dissipation rate in the gas.

The mechanism postulated above is consistent with qualitative observations of slugging behaviour, in that it is only when a solid slug is finally extinguished, and the raining down process (and hence the downward-acting force) ceases, that a new solid slug can form and start to move up the tube. Energy that the entering fluid, in the absence of this downward-acting force, would transfer to the bottom zone in the form of kinetic and potential energy is instead immediately dissipated. On this basis (given that the bottom, fixed zone is of the same average size as the upper, slugging zone) the implication of the proposed mechanism is that the excess pressure drop that is a consequence of slug formation can be largely attributed to the potential energy created by the solid slugs, eqn (15.10).

Multi-slug systems

The behaviour patterns outlined above make specific reference to single-slug systems. In the following section, we shall see what it is that determines the initial length of a solid slug, and hence how many of them will form in a bed of a given height. For systems containing more than a single slug the basic mechanism remain essentially the same, but gives rise to behaviour patterns complicated by a number of factors, which render fully quantitative predictions difficult to achieve. Certain features of multi-slug behaviour may, nevertheless, be inferred from the arguments presented for single-slug systems, permitting the following tentative observations.

The intermediate solid slugs, in addition to losing particles from below, receive them from above; this means that the solid slug particles are subject to contact forces transmitted downwards from the impact surface above – precisely the mechanism responsible for holding down the bottom packed zone in contact with the distributor as postulated in the single-slug analysis. This could result in some decrease in the intermediate solid slug velocities, with a consequential increase in dissipation rates within the solid slugs themselves. Another likely outcome is a reduction in the particle-shedding rate from the base of the solid slug, and hence a reduction in the fluid slug velocity: this latter phenomenon is precisely that utilized by Wallis (described in Chapter 9) in his procedure for measuring particle-phase dynamic-wave velocities from particle shedding experiments.

Fluid pressure loss due to potential energy dissipation

Experimental data for intermediate solid and fluid slug velocities, by means of which the above deductions could be tested, are unavailable. However, the fact that they simply describe the *details* of the potential energy dissipation process should have little bearing on the overall relation of fluid pressure loss to potential energy creation: eqn (15.10), applied to any slugging bed for which L_B represents its height under incipient fluidization conditions, should represent a reasonable approximation of this phenomenon regardless of the number of solid slugs into which L_B is divided, or even whether the fluid slugs are square-nosed, round-nosed or asymmetric:

$$\Delta P_{PE} = L_B(1 - \varepsilon_{mf})(\rho_p - \rho_f)gL_A f/2U_0. \tag{15.14}$$

Irregular oscillatory behaviour

For reasons that will shortly become apparent, it is extremely unlikely that in a multi-slug system all solid slugs would start out having the same length: And even if they did, any variations in velocity and particle-shedding rates, occurring for reasons discussed above, would certainly lead to irregular bed surface oscillation characteristics: L_A and f in eqn (15.14) come to represent average values for multi-slug systems. It will be seen that this irregularity is reflected in both experimentally determined bed surface elevation and fluid pressure loss cycles, and contrasts with the regular periodic behaviour displayed by single-slug systems.

Particle–particle and particle–wall frictional effects

The arguments employed in the previous section disregard completely solid frictional interactions. These, however, are known to play significant roles in certain aspects of slugging behaviour (Thiel and Potter, 1977), which we will now examine.

Particle–particle interactions: the initial length of a solid slug

It is the particle–particle frictional properties, in particular the '*angle of internal friction*' φ of the powder, which determine the initial length L_{SS} of a solid slug. A number of standard methods for measuring φ are reported by Zenz and Othmer (1960), two of them relating directly to solid slug formation.

The piston test

This method for measuring φ illustrates very clearly why it is that the length of a solid slug cannot exceed a certain predictable value, and why, given sufficient powder in the system, slugs of this maximum permissible length are formed at the onset of slugging fluidization.

The method involves a transparent cylindrical tube fitted with a piston (Figure 15.3). The piston must be able to move freely up and down, and yet be of sufficiently tight clearance to prevent any of the powder being tested entering between it and the tube wall.

Powder is then poured continuously into the cylinder. At first the piston remains free to move, but when the powder bed reaches a critical height L_{max} it suddenly becomes effectively impossible to move it upwards: a too vigorous attempt to do so could well result in the tube fracturing. The angle of internal friction for the powder is obtained from the ratio of this critical height to the tube diameter, as shown in Figure 15.3: $\tan\varphi = L_{max}/D$.

If the bed height at incipient fluidization is less than $D \tan\varphi$, then a single solid slug is formed from all the bed particles – less the small portion at the base, which remains in contact with the distributor. It is clear from the piston experiment, however, that for greater initial bed heights a single solid slug cannot be formed; instead, the top bed section, of length approaching $D \tan\varphi$, becomes the first solid slug, followed by others of this length, and finally a shorter one composed of the remaining available particles. This behaviour provides another of the methods

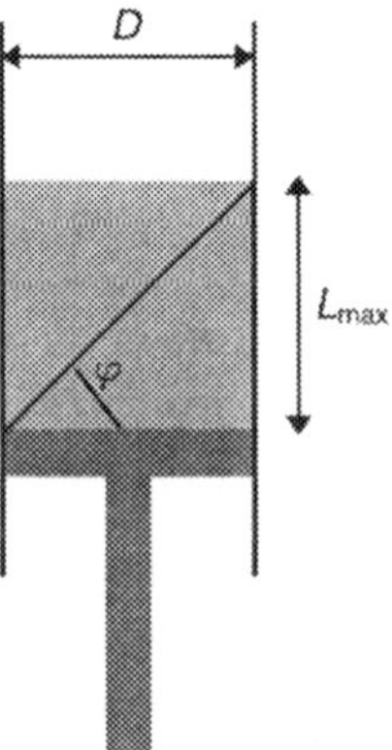

Figure 15.3 The piston test for determining the angle of internal friction.

reported by Zenz and Othmer (1960) for measuring the angle of internal friction: simply fluidize enough of the powder in a tube of suitable dimensions, and measure the initial length of the upper solid slugs.

Particle–wall interactions

In the following section we report experimental results regarding the effect of tube roughness on slugging characteristics. It turns out that the solid slug velocity is affected little, if at all, by the condition of the tube wall, whereas square-nosed fluid slug velocities *increase markedly* with tube roughness. With hindsight, it is easy to see why this should be the case. The tube wall imposes a downward force on the particles with which it is in contact; the rougher the wall the greater the downward force. Because the solid slug is densely packed, it is only particles at its bottom interface, which also represents the upper interface of the fluid slug, that are immediately affected; these fall away faster than they would in a smooth-walled tube, enabling their neighbours to do likewise and thereby leading to an increase in fluid slug velocity (Figure 15.4).

Experimental determinations of slugging characteristics

This section summarizes the essential findings of an experimental investigation into a number of the phenomena described above, in particular that of excess pressure loss under slugging conditions. Details of the equipment used, experimental procedure, range of systems studied and further experimental results are reported by Chen *et al.* (1997).

Figure 15.4 Effect of tube roughness on the fluid slug velocity.

Excess fluid pressure loss in slugging beds: effect of particle–wall and particle–particle frictional interactions

The effect of *wall roughness* was investigated in three similar air-fluidized columns, two of which were lined with sandpaper, which provided surface irregularities of some 0.1 mm and 0.5 mm. A narrow vertical strip was left uncovered in the lined columns to permit visual observation of the bed. Tests were performed using three batches of approximately spherical sand particles having mean diameters of 0.34 mm, 0.45 mm and 0.8 mm. The pressure immediately above the air distributor was measured continuously by means of a piezoelectric transducer, and logged into a microcomputer where the mean pressure loss ΔP and the slugging frequency f were evaluated. Video camera recordings of the bed surface position enabled its extent of oscillation L_A to be determined for each run. All tests were carried out at a range of initial bed heights corresponding to somewhat over 1 up to 2.5 'maximum length' solid slugs ($L_{SS} = D \tan \varphi$), at gas flow rates of up to about $3u_{mf}$.

The ratios of the *kinetic to potential energy requirements*, evaluated by means of eqn (15.11), showed clearly that the potential energy term dominated overwhelmingly for every run.

Square-nosed slugging systems

Figures 15.5–15.7 all relate to one square-nosed slugging system (0.8 mm sand, initial bed height 0.7 m) fluidized in both the smooth- and a rough-walled tube. These results are typical of all the square-nosed systems tested. The measured pressure drops are shown in Figure 15.5 as functions of gas flux U_0. It is clear that the progressive increase in ΔP with increasing U_0 beyond U_{mf} is quite independent of wall roughness.

The fact that gas pressure loss is unaffected by wall roughness could be thought to indicate that wall conditions play no part in determining slugging characteristics. This would be quite wrong, as is clear from Figures 15.6, which reports bed surface displacements L_A and slugging frequencies f as functions of gas flux for both the smooth- and rough-walled systems. These observations reveal the significant influence of the wall condition over the full reported range, the rough wall leading to consistently lower surface displacements and higher oscillation frequencies.

The qualitative reason for these differences in behaviour for the smooth- and the rough-walled tubes was anticipated in the closing

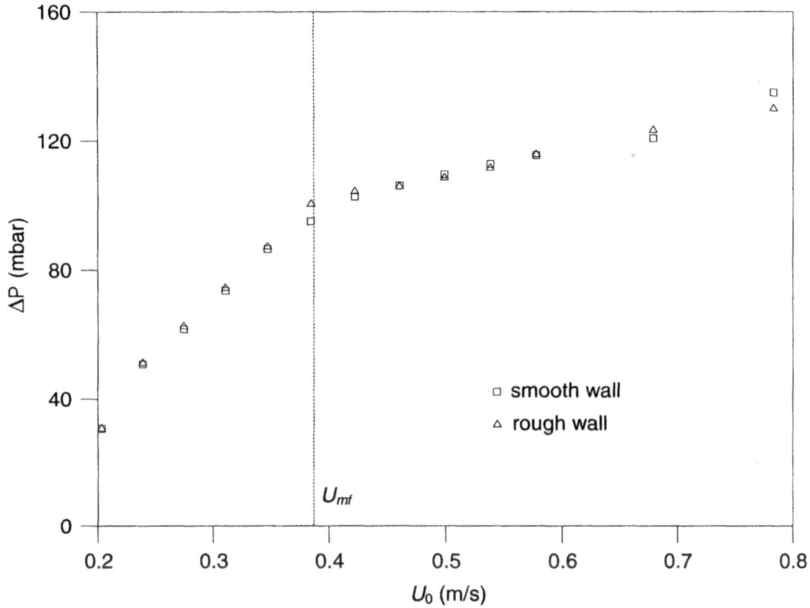

Figure 15.5 Illustrative example of observed pressure loss in slugging beds.

paragraph of the previous section. It is a consequence of the increased shedding rate of particles from the base of solid slugs in rough-walled systems – which amounts to an increase in gas slug velocity; as the solid slug is consumed faster, there is a decrease in bed surface displacement and an increase in slugging frequency.

The expression for excess pressure loss due to potential energy dissipation, eqn (15.14), contains the *product* of L_A and f. Figure 15.7 shows this

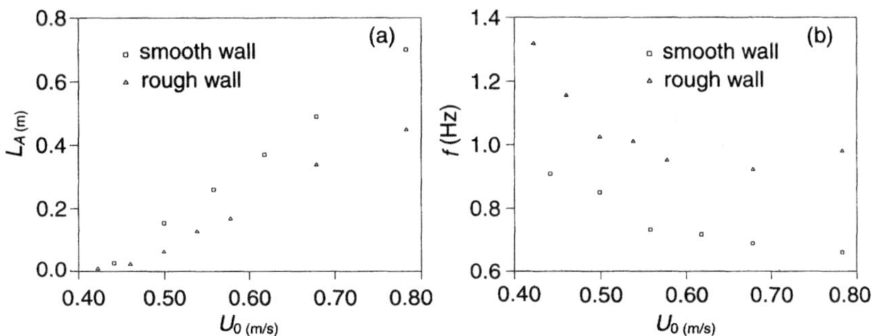

Figure 15.6 Slugging characteristics for illustrative example. (a) Bed surface displacement. (b) Slugging frequencies.

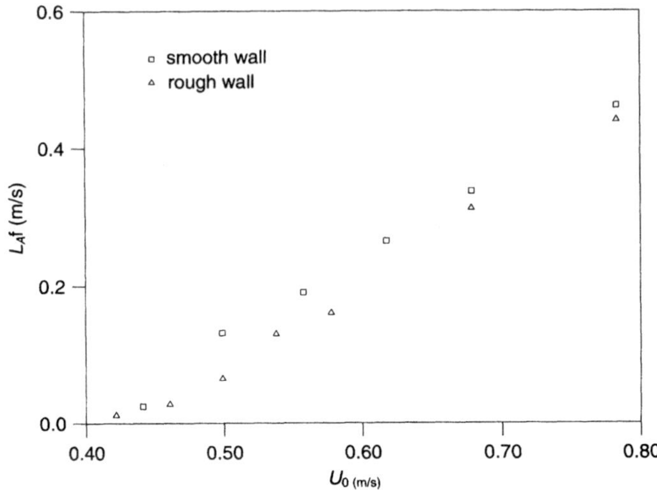

Figure 15.7 The product of bed surface displacement and slugging frequency for the illustrative example.

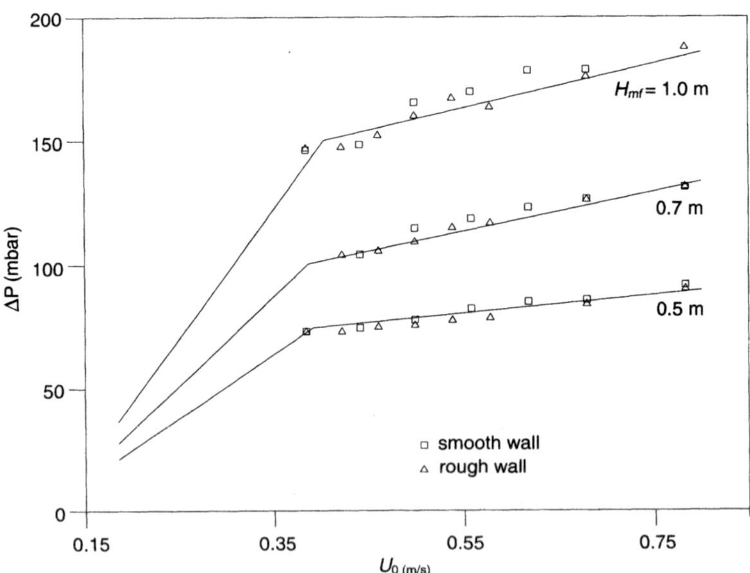

Figure 15.8 Pressure drop results for illustrative example (middle set) and same system at lower and higher particle loadings. Continuous lines, best fit through all measured pressure drop data points (from Figure 15.5 for illustrative example); points, calculated from eqn (15.14) using measured L_A and f values.

product as a function of gas flux for the illustrative example: there is very little to chose between the smooth- and rough-walled system results this time, the points effectively forming a single curve over the full operating range. This is to be expected if the sole effect of wall roughness is to increase the gas slug velocity: this would lead to both the extent L_A and the period $(1/f)$ of the bed surface oscillations being affected in the same proportion, so that the product, $L_A f$, remains unchanged.

The implication of these results is that the excess pressure loss that results from the potential energy requirement for slugging is independent of tube roughness. Figure 15.8 compares measured pressured loss for the illustrative example (and also for two other bed loadings, one lower and the other higher, of the same particles in the same tubes) with estimates based on eqn (15.14). Agreement is excellent, as was also found to be the case for all particle systems tested at all bed loadings.

Round-nosed, axisymmetric slugging systems

A counterpart series of experiments was performed to determine the effect of tube roughness for these systems. The results for excess pressure loss were found to be effectively identical to those for square-nosed systems: eqn (15.14) was found to relate excess pressure loss to the slugging characteristics L_A and f for all systems tested, precisely as illustrated for square-nosed systems in Figure 15.8. Regardless of this, however, the mechanism described above for square-nosed slugging (involving solely increased particle shedding rates induced by rough tube walls) certainly did not apply. In fact, the bed surface displacement was sometimes observed to *increase* with tube roughness, with a corresponding decrease in oscillation frequency. This could be attributed to the fact that particle shedding in round-nosed systems occurs largely by convection at the wall, and could therefore be impeded by increases in surface roughness.

Particle–particle frictional interactions

The effect of particle–particle frictional interactions on excess pressure loss was investigated in the same way as described above. Smooth tubes were used, with beds of polished glass spheres and sand particles having essentially the same size distribution and density, but differing in their particle–particle frictional properties. Although the slugging characteristics, L_A and f, differed for the two particle species, the product $L_A f$ remained the same and, once again, excess pressure loss was well accounted for by eqn (15.14). This reinforces the conclusions reached above with respect to the excess pressure loss in smooth- and rough-walled tubes for both square- and round-nosed slugging beds: regardless of the precise mechanism responsible for the dissipation process, which in any case clearly differs according to slug type, the potential energy requirement for slugging represents the dominant contribution to excess pressure loss, able to account quantitatively for this effect in all the systems tested in this study.

Instantaneous pressure and bed surface displacement fluctuations

By displaying the bed pressure drop continuously on an oscilloscope screen mounted alongside the bed, it was possible to video record the bed surface position together with the corresponding values of ΔP across

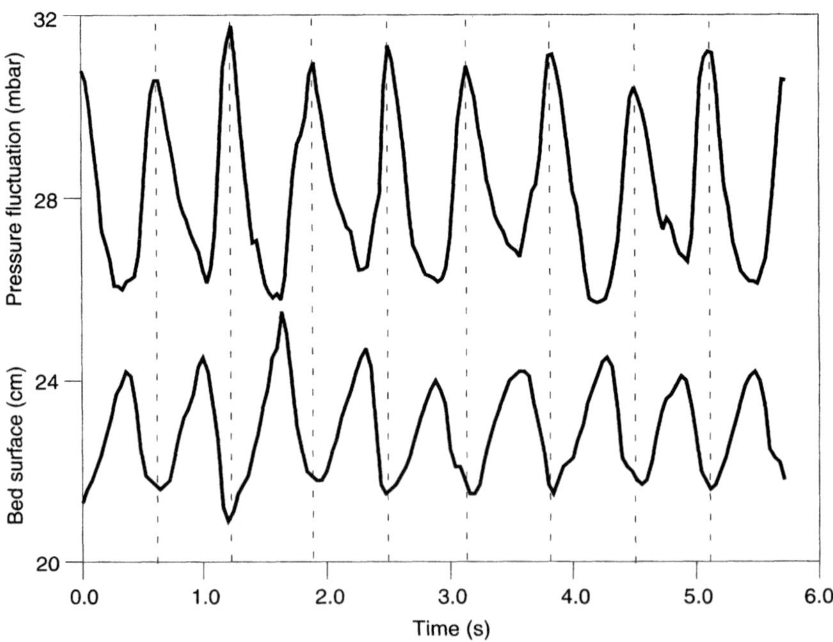

Figure 15.9 Fluid pressure and bed surface oscillations for a single square-nosed slugging system.

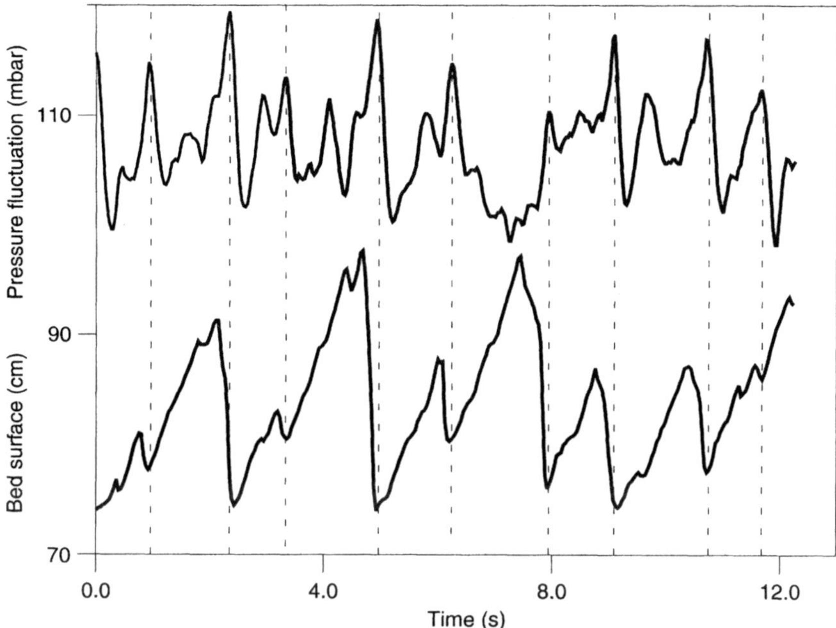

Figure 15.10 Fluid pressure and bed surface oscillations for a multiple square-nosed slugging system.

a slugging cycle. A typical example, for a single, square-nosed slugging system, is shown in Figure 15.9: each of the two oscillatory records consists effectively of a single harmonic.

These observed oscillations are much more symmetrical than would be inferred from the idealized picture of square-nosed slugging depicted in Figure 15.1. Inertial effects could be in part responsible for smoothing the abrupt changes postulated for the birth and death of a solid slug. What the results clearly show, however, is that the pressure loss approaches a maximum at the point in the slugging cycle where the bed surface height is approaching a minimum – corresponding to a maximum height static zone at the distributor where, it has been argued, the major excess energy dissipation occurs.

Finally, Figure 15.10 shows pressure drop and bed surface oscillations for an initially taller bed, which leads to multiple slug behaviour. Although, as anticipated earlier, the behaviour is less regular, it is still possible to see that a maximum in pressure loss clearly corresponds to a minimum in bed surface height.

References

Baker, C.G.J. and Geldart, D. (1978). An investigation into the slugging characteristics of large particles. *Powder Technol.*, **19**, 177.

Chen, Z., Gibilaro, L.G. and Foscolo, P.U. (1997). Fluid pressure loss in slugging fluidized beds. *Chem. Eng. Sci.*, **52**, 55.

Gibilaro, L.G., Foscolo, P.U., Rapagnà, S. *et al.* (1998). Particle shedding in slugging fluidized beds. In: *Fluidization IX* (L.-S. Fan and T.M. Knowlton, eds). Engineering Foundation.

Grace, J.R. and Harrison, D. (1970). Design of fluidized beds with internal baffles. *Chem. Proc. Eng.*, **46**, 127.

Hovmand, S. and Davidson, J.F. (1971). Pilot plant and laboratory scale fluidized reactors; the relevance of slug flow. In: *Fluidization* (J.F. Davidson and D. Harrison, eds). Academic Press.

Jeffrey, A. (1976). *Quasilinear Hyperbolic Systems and Waves.* Pitman.

Thiel, W.J. and Potter, O.E. (1977). Slugging in fluidized beds. *Ind. Eng. Chem. Fund.*, **16**, 242.

Toomey, R.D. and Johnstone, H.F. (1952). Gaseous fluidization of solid particles. *Chem. Eng. Prog.*, **48**, 220.

Zenz, F.A. and Othmer, D.F. (1960). *Fluidization and Fluid–Particle Systems.* Reinhold Publishing Corporation.

16

Two-dimensional simulation

The two-dimensional, two-phase particle bed model

The analyses presented in previous chapters have been in terms of the *one-dimensional* equations of change. Only the scaling relations, derived in Chapter 13 on the basis of these equations, can be claimed to relate strictly to full three-dimensional bed behaviour. In addition to this imposed limitation, most of the conclusions reached up to this point have been based on an effectively *single-phase* formulation: all gas-fluidized beds and liquid-fluidized beds of moderate to high density particles are dominated in their behaviour by the particle-phase mass and momentum relations, thereby supporting this second simplification. The predictive ability of analyses that start out from these assumptions has featured prominently in this book.

It is clear, however, that a complete study of the development of bubbles and other inhomogeneities in fluidized beds requires multi-dimensional analysis. There are considerable incentives for such work, and, following the pioneering 'super-computer' modelling approach of Gidaspow *et al.* (1986), a number of numerical treatments has appeared in the literature. More recently, the *fluidization quality* issue has been confronted directly by numerical integration of the two-dimensional equations of change (Anderson *et al.*, 1995); two sets of parameter values were employed, corresponding to an air-fluidized and a water-fluidized system (both unstable); it was found that, in conformity with qualitative experimental evidence, only the former gave rise to bubble-like structures. A subsequent, related study (Glasser *et al.*, 1997) also employed direct numerical integration to trace the evolution of one- and two-dimensional waves in fluidized beds for a wide range of parameter values; this led to a tentative criterion distinguishing bubbling from non-bubbling heterogeneous systems.

The sections that now follow report a two-dimensional, two-phase formulation of the *particle bed* model (Chen *et al.*, 1999), and apply it to a number of key problems in fluidization dynamics. The generalization follows naturally from the one-dimensional, two-phase treatment described in Chapter 11. It remains fully predictive, no arbitrary or adjustable parameters being introduced.

The two-dimensional force interactions

The one-dimensional *particle bed* model has been formulated in terms of the *primary* fluid–particle interaction forces, which alone may be considered to support a fluidized particle under steady-state *equilibrium* conditions, together with *particle-phase elasticity*, which provides a force proportional to void fraction gradient (or particle concentration gradient) and so comes into play under *non-equilibrium* conditions. Only axial components of these interactions have been considered so far. Generalizing these considerations to encompass lateral force components is a straightforward matter, but, as we now see, calls for some modification in the constitutive relation for drag in order to unify the axial and lateral constitutive expressions. The following derivations are expressed in terms of volumetric particle concentration α rather than void fraction ε: $\alpha = 1 - \varepsilon$.

The primary forces acting on a fluidized particle

Equilibrium conditions

Under equilibrium conditions only axial forces come into play. The primary axial (z-direction) forces acting on a fluidized particle comprise: gravity f_{gz}, buoyancy f_{bz} (the net effect of the mean, axial fluid pressure gradient in the bed), and drag f_{dz}. The sum of the gravity and buoyancy forces is:

$$f_{gz} + f_{bz} = -\frac{\pi d_p^3}{6}\left(\rho_p g + \frac{\partial p}{\partial z}\right). \tag{16.1}$$

Under equilibrium conditions, for which the fluid pressure gradient is given by

$$\frac{\partial p}{\partial z} = -(\rho_p \alpha + \rho_f(1-\alpha))g, \tag{16.2}$$

eqn (16.1) becomes:

$$f_{gz} + f_{bz} = -\frac{\pi d_p^3}{6}(\rho_p - \rho_f)(1-\alpha)g. \tag{16.3}$$

The drag force f_{dz} may be obtained by extension of the drag coefficient relation for the drag f_{d0} experienced by a solitary, *unhindered* particle subjected to a steady fluid velocity U_0:

$$f_{d0} = C_D \cdot \frac{\rho_f U_0^2}{2} \cdot \frac{\pi d_p^2}{4}, \tag{16.4}$$

where the drag coefficient C_D may be expressed by the empirical Dallavalle relation:

$$C_D = \left(0.63 + \frac{4.8}{Re^{0.5}}\right)^2, \quad Re = \frac{\rho_f U_0 d_p}{\mu_f}. \tag{16.5}$$

For the case of a *stationary fluidized particle* in equilibrium in a bed with particle concentration α, the drag coefficient $C_{D\alpha}$ relates to the unhindered value C_D through:

$$C_{D\alpha} = C_D(1-\alpha)^{-3.8}. \tag{16.6}$$

This relation merely expresses the hypothesis of common void fraction dependency of particle drag for all flow regimes, which led to eqn (4.25). Inserting it in eqn (16.4) yields:

$$f_{dz} = C_D \frac{\rho_f U_0^2}{2} \cdot \frac{\pi d_p^2}{4} (1-\alpha)^{-3.8}. \tag{16.7}$$

This expression for axial drag differs from that employed in previous chapters, namely:

$$f_{dz} = \frac{\pi d_p^3}{6} \cdot (\rho_p - \rho_f) g \left(\frac{U_0}{u_t}\right)^{\frac{4.8}{n}} \cdot (1-\alpha)^{-3.8}. \tag{16.8}$$

For the limiting extremes of the viscous regime ($n = 4.8$, u_t given by eqn (2.12), $C_D = 24/Re = 24\mu_f/\rho_f U_0 d_p$) and the inertial regime ($n = 2.4$, u_t given by eqn (2.13), $C_D = 0.44$) these two expressions for axial drag, eqns (16.7) and (16.8), become identical. However, they interpolate between the two limits somewhat differently (Gibilaro *et al.*, 1985). The advantage of the form used up to now, eqn (16.8), is that by incorporating the empirical parameter n it delivers the Richardson–Zaki relation, $U_0 = u_t \varepsilon^n$, under equilibrium conditions, $f_{z0} = 0$, thereby ensuring an accurate interpolation in the intermediate flow regime. On the other hand, the form of eqn (16.7) renders it relevant for all flow directions, and therefore more appropriate for multi-dimensional applications.

On this basis, the total primary axial force f_{z0} for a bed in *equilibrium* becomes:

$$f_{z0} = \frac{\pi d_p^3}{6} \left(C_D \frac{3\rho_f U_0^2}{4 d_p} \right) \cdot (1-\alpha)^{-3.8} - (\rho_p - \rho_f)(1-\alpha) g = 0. \tag{16.9}$$

In the following section, we make use of this equilibrium relation to provide an estimate for particle-phase elasticity.

General, non-equilibrium conditions: the net primary axial force

For the general two-dimensional situation, where the particle has velocity $\mathbf{u}_\mathrm{p}$ (axial and lateral components u_p and v_p respectively) the above relations must be expressed in terms of the relative fluid/particle axial velocity:

$$f_{dz} = C_D \frac{\rho_f(1-\alpha)(u_f - u_p)|(1-\alpha)(\mathbf{u}_\mathrm{f} - \mathbf{u}_\mathrm{p})|}{2} \cdot \frac{\pi d_p^2}{4}(1-\alpha)^{-3.8}$$

$$= \frac{\pi d_p^3}{6} \cdot C_D \frac{3\rho_f(u_f - u_p)|(\mathbf{u}_\mathrm{f} - \mathbf{u}_\mathrm{p})|}{4d_p} \cdot (1-\alpha)^{-1.8}, \qquad (16.10)$$

where C_D is evaluated from eqn (16.5) with:

$$Re = \frac{\rho_f(1-\alpha)|(\mathbf{u}_\mathrm{f} - \mathbf{u}_\mathrm{p})|d_p}{\mu_f}. \qquad (16.11)$$

The net primary axial force, $f_z = f_{gz} + f_{bz} + f_{dz}$, is thus obtained from eqns (16.1) and (16.10):

$$f_z = \frac{\pi d_p^3}{6}\left(C_D \frac{3\rho_f(u_f - u_p)|(\mathbf{u}_\mathrm{f} - \mathbf{u}_\mathrm{p})|}{4d_p} \cdot (1-\alpha)^{-1.8} - \rho_p g - \frac{\partial p}{\partial z}\right). \qquad (16.12)$$

General, non-equilibrium conditions: the net primary lateral force

The fluid pressure gradient force on a particle in the x-direction (analogous to buoyancy in the z direction) is given by:

$$f_{bx} = -\frac{\pi d_p^3}{6}\frac{\partial p}{\partial x}. \qquad (16.13)$$

Lateral drag, the only other primary force we need consider, may be expressed in the same way as for the axial component, so that the net primary lateral force becomes:

$$f_x = f_{dx} + f_{bx} = \frac{\pi d_p^3}{6}\left(C_D \cdot \frac{3\rho_f(v_f - v_p) \cdot |(\mathbf{u}_\mathrm{f} - \mathbf{u}_\mathrm{p})|}{4d_p} \cdot (1-\alpha)^{-1.8} - \frac{\partial p}{\partial x}\right), \qquad (16.14)$$

where C_D is obtained from eqn (16.5), with Re given by eqn (16.11) as before.

This completes the assembly of the primary force interactions for the two-dimensional formulation. In order to arrive at the two-dimensional counterpart of the *particle bed* model, it only remains to consider the effect of particle-phase elasticity.

Fluid-dynamic elasticity of the particle phase

Axial component

The concept of fluid-dynamic elasticity of the particle phase was introduced in Chapter 8. The arguments leading to eqn (8.12) apply directly to the axial component of the two-dimensional case now under consideration; the additional force $f_{\Delta z}$ on a fluidized particle, which comes into play as a result of a particle concentration gradient in the z direction, is:

$$f_{\Delta z} = -\left(\frac{2d_p}{3} \cdot \frac{\partial f_{z0}}{\partial \alpha}\right) \cdot \frac{\partial \alpha}{\partial z}, \tag{16.15}$$

where f_{z0}, the expression for the net primary force on a fluidized particle under equilibrium conditions, is given by eqn (16.9). The fact that this expression differs from the one used previously in Chapter 8 makes no difference to its derivative $\partial f_{z0}/\partial\alpha$ evaluated at equilibrium ($f_{z0} = 0$), which is only influenced by the dependence of drag on α – which is the same in the two alternative formulations. Thus we obtain, as before:

$$f_{\Delta z} = -\frac{\pi d_p^3}{6} 3.2 g d_p(\rho_p - \rho_f)\frac{\partial \alpha}{\partial z}. \tag{16.16}$$

Lateral component

Under conditions of equilibrium, lateral flow rates and the lateral pressure gradient in the fluid are all zero, so that there are no non-zero force components acting in the x-direction. The additional force $f_{\Delta x}$, obtained by the lateral counterpart of eqn (16.15), is thus also zero: there is no effective elasticity in the x-direction.

Particle- and fluid-phase force components

The above force interactions have been expressed in terms of the forces f acting on a single fluidized particle: f_z, f_x and $f_{\Delta z}$. Forces F acting on the particle phase are obtained by multiplying f by the number of particles per unit volume of suspension:

$$F = \frac{6\alpha}{\pi d_p^3} \cdot f. \tag{16.17}$$

Particle-phase force components

The total force acting on a single particle is the sum of the net primary force and the force resulting from particle-phase elasticity. The axial component F_{pz} is thus obtained from eqns (16.12), (16.16) and (16.17):

$$F_{pz} = C_D \frac{3\alpha\rho_f(u_f - u_p) \cdot |(\mathbf{u}_\mathrm{f} - \mathbf{u}_\mathrm{p})|}{4d_p} \cdot (1-\alpha)^{-1.8} - \alpha\rho_p g - \alpha\frac{\partial p}{\partial z}$$
$$- 3.2gd_p\alpha(\rho_p - \rho_f)\frac{\partial \alpha}{\partial z}. \tag{16.18}$$

Similarly, the lateral component F_{px} follows from eqns (16.14) and (16.17):

$$F_{px} = C_D \cdot \frac{3\alpha\rho_f(v_f - v_p) \cdot |(\mathbf{u}_\mathrm{f} - \mathbf{u}_\mathrm{p})|}{4d_p} \cdot (1-\alpha)^{-1.8} - \alpha\frac{\partial p}{\partial x}. \tag{16.19}$$

Fluid-phase force components

The fluid-phase forces are readily obtained from the particle-phase relations for fluid–particle interaction (drag and the pressure gradient force), which acts in the opposite direction on the fluid, together with gravity and the effect of the fluid pressure gradient across the control volume; fluid viscosity effects are considered only in so far as they as they contribute to fluid–particle drag. Thus we have for the axial component F_{fz}:

$$F_{fz} = -C_D \frac{3\alpha\rho_f(u_f - u_p) \cdot |(\mathbf{u}_\mathrm{f} - \mathbf{u}_\mathrm{p})|}{4d_p} \cdot (1-\alpha)^{-1.8}$$
$$- (1-\alpha)\rho_f g - (1-\alpha)\frac{\partial p}{\partial z}; \tag{16.20}$$

and for the lateral component F_{fx}:

$$F_{fx} = -F_{px}. \tag{16.21}$$

The two-dimensional, two-phase equations of change

The one-dimensional, two-phase equations presented in Chapter 8, eqns (8.21)–(8.24), are readily generalized to multi-dimensional form.

Particle phase

$$\frac{\partial \alpha}{\partial t} + \nabla \cdot (\alpha \mathbf{u}_{\mathrm{p}}) = 0, \quad \textit{Continuity} \tag{16.22}$$

$$\rho_p \alpha \frac{\partial \mathbf{u}_{\mathrm{p}}}{\partial t} + \rho_p \alpha (\mathbf{u}_{\mathrm{p}} \cdot \nabla)\mathbf{u}_{\mathrm{p}} = \mathbf{F}_{\mathrm{p}}, \quad \textit{Momentum} \tag{16.23}$$

where the elements of the particle-phase force vector $\mathbf{F}_{\mathrm{p}}$ are given by eqns (16.18) and (16.19).

Fluid phase

$$\frac{\partial \alpha}{\partial t} - \nabla \cdot ((1-\alpha)\mathbf{u}_{\mathrm{f}}) = 0, \quad \textit{Continuity} \tag{16.24}$$

$$\rho_f (1-\alpha) \frac{\partial \mathbf{u}_{\mathrm{f}}}{\partial t} + \rho_f (1-\alpha)(\mathbf{u}_{\mathrm{f}} \cdot \nabla)\mathbf{u}_{\mathrm{f}} = \mathbf{F}_{\mathrm{f}}, \quad \textit{Momentum} \tag{16.25}$$

where the elements of the fluid-phase force vector $\mathbf{F}_{\mathrm{f}}$ are given by eqns (16.20) and (16.21).

Equations (16.18)–(16.25) fully describe the general system. It is noteworthy that the only empirical input is the single unhindered-particle drag-coefficient $C_{\hat{D}}$, eqn (16.5), and even this, under low Reynolds number conditions, takes the theoretical form: $C_{\hat{D}} = 24/Re_t$.

Boundary conditions

The following section reports numerical solutions to the above equations, which relate to a 'two-dimensional' slice of a fluidized bed of width L_x equal to 0.2 m, and of sufficient height L_z to allow for bed expansion – typically 0.4 m to accommodate an initial packed bed height of some 0.25 m. The system is restricted laterally by impermeable, rigid walls, and at the top and bottom by horizontal planes permeable for the fluid but not for the particles. Boundary conditions for each of the variables (u_f, u_p, v_f, v_p, p, and α) are imposed as follows:

$$x = 0, \quad L_x: \frac{\partial \alpha}{\partial x} = 0, \quad \frac{\partial u_f}{\partial x} = \frac{\partial u_p}{\partial x} = 0, \quad v_f = v_p = 0; \tag{16.26}$$

$$z = 0: \; v_f = v_p = u_p = 0, \quad p = p_0, \quad \alpha = \alpha_0, \quad u_f = U_0/(1-\alpha); \tag{16.27}$$

$$z = L_z: \frac{\partial \alpha}{\partial z} = 0, \quad \frac{\partial u_f}{\partial z} = \frac{\partial v_f}{\partial z} = 0, \quad u_p = v_p = 0. \tag{16.28}$$

Conditions (16.26) impose x-periodicity on the solutions; they are the same as those used by Anderson *et al.* (1995), cited earlier, where it is pointed out that they 'suppress any solutions in the form of oblique travelling waves', a limitation unlikely to be of any significance to normal fluidized bed behaviour. Condition (16.28) represents the modelling approximation, strictly applicable at $z = \infty$, which assumes the freeboard to be sufficiently large to ensure negligible particle concentration at the upper boundary, $z = L_z$.

Initial conditions

Two types of initial conditions were employed in the simulations reported below.

Initial conditions 1

The system is set at a chosen uniform, equilibrium condition, at fluid flux U_{0i} and bed height H_0, and is then simply subjected to a change in fluid flux (to U_0) at the start of the integration period. This procedure simulates closely the experimental procedures usually adopted in practice. Perturbations develop naturally as a result of the shift away from the initial equilibrium state, which is assumed to correspond to stationary particles with fluid flow solely in the vertical direction, and the fluid pressure constant in all horizontal planes, and distributed vertically to reflect suspension weight:

$$u_p = v_p = 0;$$

$$u_f = U_{0i}/(1-\alpha) = u_t(1-\alpha)^{n-1}, \quad v_f = 0;$$

$$\partial p/\partial x = 0, \; p = p_a + (H_0 - z)(\alpha_0 \rho_p + (1-\alpha_0)\rho_f)g. \tag{16.29}$$

The freeboard, included to provide the necessary volume for bed expansion, is assumed to be initially at atmospheric pressure p_a, void of particles, and subjected to fluid flow only in the vertical direction:

$$\alpha = 0, \quad u_f = U_0, \quad u_p = v_p = v_f = 0, \quad p = p_a. \tag{16.30}$$

Initial conditions 2

In this case the initial equilibrium state, computed by means of eqn (16.30), is set to correspond to the fluid flux that is maintained throughout the run ($U_0 = U_0$); in addition, a small void fraction perturbation is imposed at the bottom of the bed. In cases where this condition has been adopted, the perturbation is set to consist of three small hemispherical voids spaced evenly across the bed, as shown in Figure 16.1. This arrangement may be thought to simulate an imperfect distributor. The perturbations are computed by means of the following expression:

$$\alpha = \alpha_0 - \alpha\left(1 - \frac{(x - x_0)^2 + z^2}{0.02^2}\right), \text{ for } \frac{(x - x_0)^2 + z^2}{0.02^2} \leq 1;$$

$$\alpha = \alpha_0, \text{ for } \frac{(x - x_0)^2 + z^2}{0.02^2} > 1;$$

$$x_0 = 0.05, 0.10, 0.15. \tag{16.31}$$

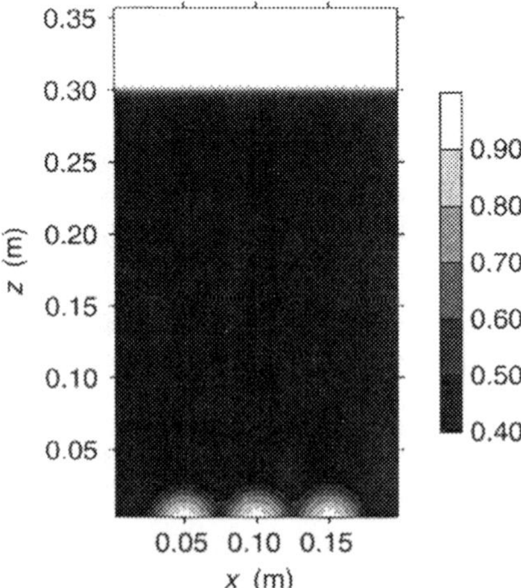

Figure 16.1 Initial condition for a distributor-induced void fraction perturbation.

The scale to the left of Figure 16.1 is applicable to all the pictorial representations of void fraction distribution presented in the following section.

Numerical simulations

The reader is referred to Chen *et al.* (1999) for an outline of the numerical procedure adopted for the computation of the three sets of results that now follow; the first two of these repeat essentially those reported in that publication. The final set provides a more detailed examination of the quantitative possibilities offered by multi-dimensional simulation of the fluidized state.

Unsteady-state contraction and expansion of homogeneous, liquid-fluidized beds

The response of homogeneous, liquid-fluidized beds to sudden changes in liquid flux was analysed in Chapter 5 on the sole basis of equilibrium, mass-conservation considerations. It was found that for step reductions in fluid flux, from U_1 to U_2, the bed surface is predicted to fall linearly with time, at velocity $U_2 - U_1$, to the new equilibrium level. At the same time a concentration shock-wave travels upwards from the distributor towards the descending bed surface, at the velocity given by eqn (5.8); when these two interfaces meet, the transient response period is completed. This simple behaviour is well known to be in excellent quantitative agreement with experimental observations. It therefore provides a good initial test of the model formulation, described in the previous section, and also of the numerical code employed to implement it.

For these simulations *initial conditions* 1, eqns (16.29) and (16.30), were employed, the chosen equilibrium state corresponding to the initial steady state of the bed before the fluid flux change is imposed.

The contracting bed

Figure 16.2 shows results for an example simulation of water fluidization of soda-glass spheres. Agreement with the idealized response, derived in Chapter 5, is excellent both qualitatively, as is immediately apparent from the figures, and quantitatively: a simulated bed surface fall velocity of 80 mm/s as against the predicted value ($U_2 - U_1$) of 81 mm/s; and a simulated velocity for the interface separating the two equilibrium zones

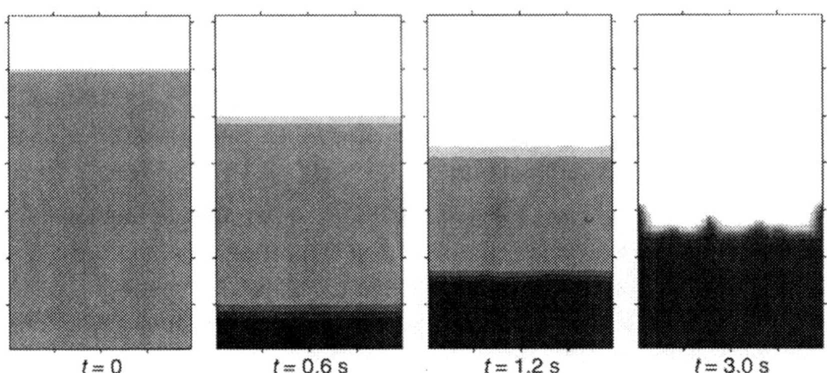

Figure 16.2 Response of a liquid-fluidized bed to a sudden decrease in fluid flux. System: ambient water/soda-glass spheres: $\rho_f = 1000\,\text{kg/m}^3$, $\mu_f = 0.001\,\text{Ns/m}^2$, $\rho_p = 2500\,\text{kg/m}^3$, $d_p = 2\,\text{mm}$, $U_1 = 113\,\text{mm/s}$, $\varepsilon_1 = 0.75$, $U_2 = 32\,\text{mm/s}$, $\varepsilon_2 = 0.45$.

of 65 mm/s as against the predicted kinematic-shock value of 68 mm/s. The response remains effectively one-dimensional throughout the transient response period.

The expanding bed

Figure 16.3 shows what happens when, for the same bed, the flow rate change is reversed, leading to bed expansion back to its original condition. In this case the upper zone of the bed, during the transient period, is at a higher density than the lower zone, leading to the possibility of

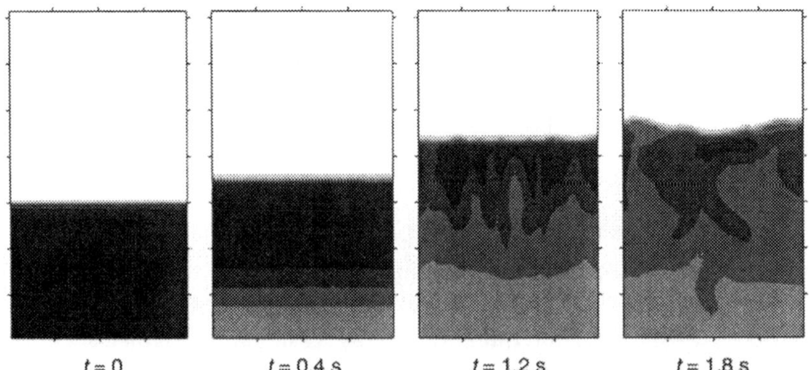

Figure 16.3 Response of a liquid-fluidized bed to a sudden increase in fluid flux. System: same as for Figure 16.2, with fluid flux change in the reverse direction.

gravitational instabilities. This phenomenon was discussed in Chapter 5 and illustrated in Figure 5.6; it has long been cited to explain the failure of the simple law, obeyed throughout by contracting beds, to apply beyond the initial stages of bed expansion. Figure 16.3 provides unequivocal support for this explanation: after some half-second of expansion, fingers of low density suspension are to be seen starting to penetrate the upper zone, reaching and disrupting the surface after about 1.2 s. For the first half-second of the expansion the behaviour remains in good quantitative agreement with the simple theory: a surface rise velocity of 79 mm/s (theoretical value: 82 mm/s), and an internal interface velocity of 160 mm/s (theoretical kinematic-shock velocity: 150 mm/s). Beyond this first half-second the expansion ceases to exhibit one-dimensional characteristics, displaying instead marked inhomogeneities that slowly fade as the final equilibrium condition is approached.

The response to distributor-induced perturbations

For these simulations *initial conditions* 2 have been employed. For each case, the initial equilibrium state has been set to correspond to a void fraction of 0.45 ($\alpha_0 = 0.55$), with the superimposed perturbation computed by means of relation (16.31) – Figure 16.1.

Stable systems

Figure 16.4 shows two stable responses to the imposed perturbation. The first represents a fine powder air-fluidized system, which exhibits a transition from homogeneous to heterogeneous behaviour at a void fraction of 0.53. At the chosen condition of $\varepsilon = 0.45$ the system is stable, as is clearly demonstrated by the simulated response: the imposed voids simply detach from the base, lose their sharp boundaries, rise through the bed as 'mushroom-shaped parvoids' of the type first observed by Hassett (1961) in liquid-fluidized systems, and penetrate the bed surface causing very little disruption; thereafter the bed returns to the homogeneous condition.

The second simulation, shown in Figure 16.4, is of a stable water-fluidized system. Initially the response is very similar to that of the previous homogeneous gas bed: mushroom-shaped parvoids that travel smoothly upwards, causing little disruption to the bed surface. This time, however, in contrast to the gas bed, there is a spontaneous formation of

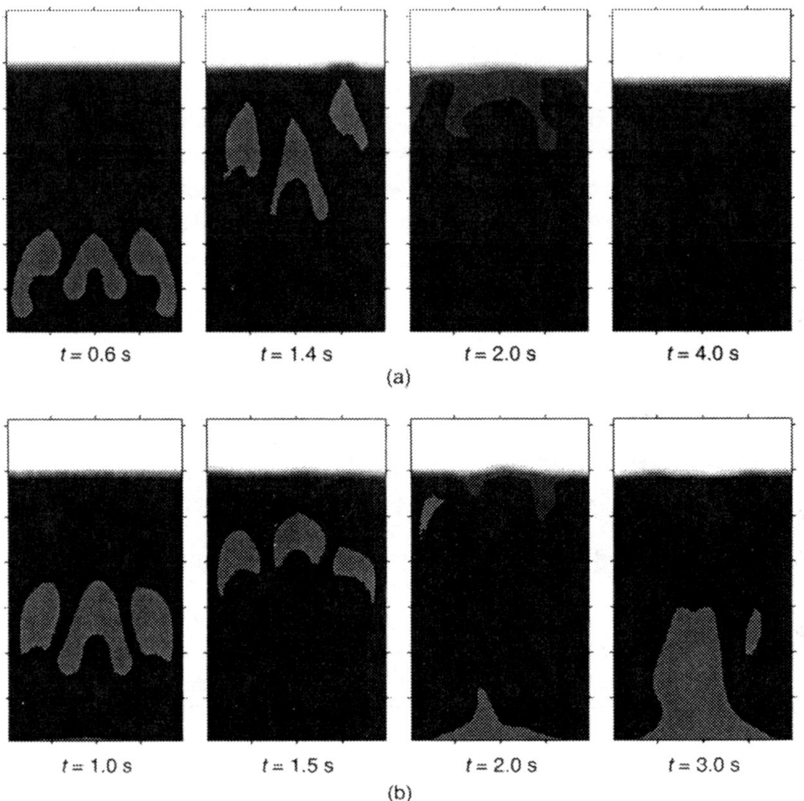

Figure 16.4 Simulation of stable, homogeneous systems.
(a) Air fluidization of fine alumina particles: $d_p = 60\,\mu m$, $\rho_p = 1500\,kg/m^3$, $u_t = 0.141$ m/s, $n = 4.68$, $\varepsilon_{mb} = 0.53$, $\varepsilon_0 = 0.45$.
(b) Water fluidization of soda-glass particles: $d_p = 2000\,\mu m$, $\rho_p = 2500\,kg/m^3$, $u_t = 0.232$ m/s, $n = 2.49$, $\varepsilon_{mb} = 0.77$, $\varepsilon_0 = 0.45$.

new parvoids at the distributor, which prolong the period of heterogeneous behaviour. This difference could well relate to the phenomenon of continual, upwards propagating, low void fraction bands observed in water-fluidized beds, as reported in Chapters 9 and 12.

Unstable systems

Figure 16.5 shows two bubbling, Geldart group B, gas-fluidized beds. The initial conditions are somewhat unrealistic for these cases, as the stable state at $\varepsilon = 0.45$ would be unrealizable in practice. Nevertheless, the early time behaviour of both systems provides an interesting comparison with the stable counterparts of Figure 16.4. This time, the initial imposed perturbations grow rapidly as they detach from the distributor, to be followed by a trail of spontaneously created voids, which also grow and coalesce to form the asymmetric bubbles that are characteristic of heterogeneous gas fluidization. After little more than 1 s all trace of the initial perturbation is lost, and the 'freely bubbling' condition, which gives rise to significant bed surface disruption, becomes firmly established.

The second example of Figure 16.5 shows a further increase in unstable behaviour brought about by an increase in both particle size and density:

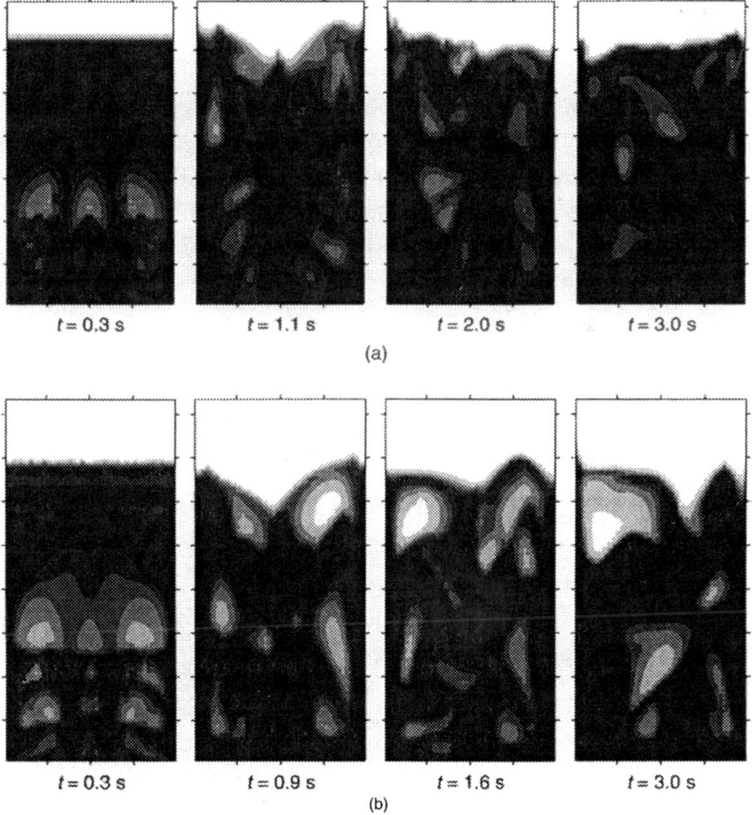

Figure 16.5 Simulation of unstable, bubbling systems.
(a) Air fluidization of alumina particles: $d_p = 380\,\mu m$, $\rho_p = 1500\,kg/m^3$, $u_t = 0.88\,m/s$, $n = 3.33$, $\varepsilon_{mb} = 0.11$ (i.e. always bubbling), $\varepsilon_0 = 0.45$.
(b) Air fluidization of sand particles: $d_p = 610\,\mu m$, $\rho_p = 2500\,kg/m^3$, $u_t = 4.06\,m/s$, $n = 2.76$, $\varepsilon_{mb} = 0.08$ (i.e. always bubbling), $\varepsilon_0 = 0.45$.

bubble growth and coalescence rates both increase, and bed surface disruption becomes extreme, with oscillation amplitudes in excess of 10 cm. The results of this simulation bear an uncanny resemblance to photographs of actual freely bubbling 'two-dimensional' gas beds, with respect to bubble shape, distribution, growth and coalescence.

Prediction of fluidization quality

As a final, more quantitative illustration of the possibilities offered by numerical simulation, we now examine two systems which have been matched in terms of the fluidization quality criteria introduced in Chapter 10. We consider first a high-temperature, high-pressure, gas-fluidized bed of supposedly unknown fluidization quality. This could represent a proposed commercial reactor.

The void fraction at minimum bubbling ε_{mb} and the perturbation-amplitude growth-rate parameter Δa are first obtained for the system from the relations given in Table 8.1 and eqn (10.7) respectively. Only the basic system properties (ρ_f, μ_f, ρ_p, d_p) are required for these evaluations. The parameters ε_{mb} and Δa locate the system on the fluidization quality map (Figure 10.7), matching it with a group D, ambient air-fluidized bed situated close to the B/D boundary. The basic properties and fluidization quality parameters of the two matched systems are given in Table 16.1.

This matching has been made on the basis of the linearized model equations. However, the very high perturbation amplitude growth rates could be thought to have strong influence on the subsequent non-linear behaviour, thereby projecting the equivalence into the non-linear, bubbling regime.

Table 16.1 Fluidization quality parameters for a proposed high temperature and pressure reactor and a matched ambient air-fluidized system

	ρ_f (kg/m^3)	μ_f (Ns/m^2 $\times$ 10^5)	ρ_p (kg/m^3)	d_p (μm)	ε_{mb}	Δa (s^{-1})
High temperature and pressure reactor	4.0	3.2	5000	450	0.08	3700
Matched ambient air system	1.2	1.8	2500	480	0.08	3800

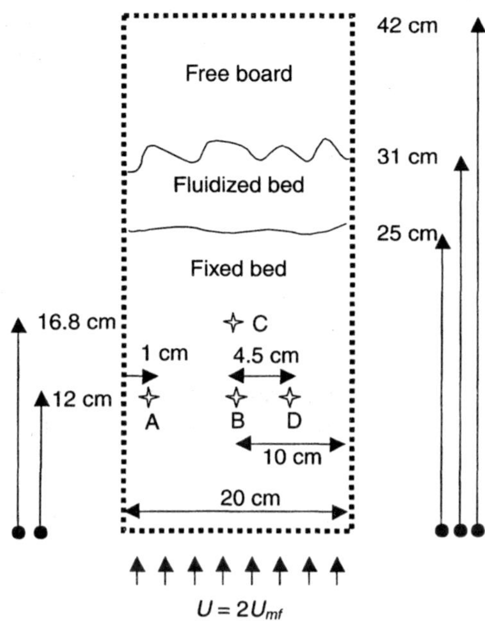

Figure 16.6 Geometric arrangement for the two-dimensional simulations.

To examine this hypothesis, the two-phase, two-dimensional *particle bed* model has been used to simulate the two matched systems (Chen *et al.*, 2001). The geometric arrangement for the simulations is shown in Figure 16.6. The gas flux was $2U_{mf}$ in both cases. Instantaneous pressure and void fraction measurements were recorded for data analysis at the points A, B, C and D. The initial conditions corresponded to beds at the point of minimum fluidization; gas rates were then set to $2U_{mf}$, resulting in the development of freely bubbling beds. The results reported below are representative of all the measured data, and confirm the equivalence of fluidization quality in the two matched beds.

Bed surface oscillations

The first comparison of the two systems shows the bed surface oscillations to be in close agreement, even in respect of the initial response immediately following the step change in gas flux (Figure 16.7). Bed surface behaviour is dominated by the size of bubbles leaving the bed, so that this equivalence supports the hypothesis that bubble size is largely

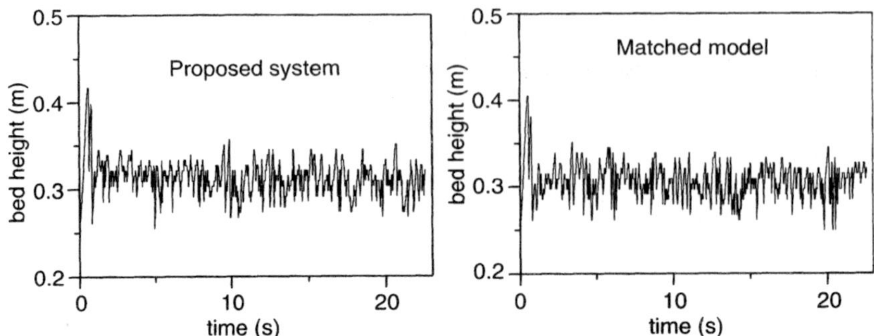

Figure 16.7 Bed surface oscillations in the two matched systems.

determined by the initial amplitude growth-rate of void fraction perturbations.

Fluid pressure fluctuations

The equivalence of the two systems is further confirmed by the fluid pressure fluctuation characteristics, reported in Figure 16.8 for position B on the bed axis. These pressures are dominated by the bed surface fluctuations reported in the previous figures. The root mean square value for the proposed high pressure and temperature system is approximately double that of the matched ambient one, reflecting the fact that the corresponding bed densities are also in this ratio.

The equivalence of these pressure fluctuations is further demonstrated by the power spectrum density functions, normalized with respect to the relevant frequency range of 0–10 Hz. These are reported in Figure 16.9,

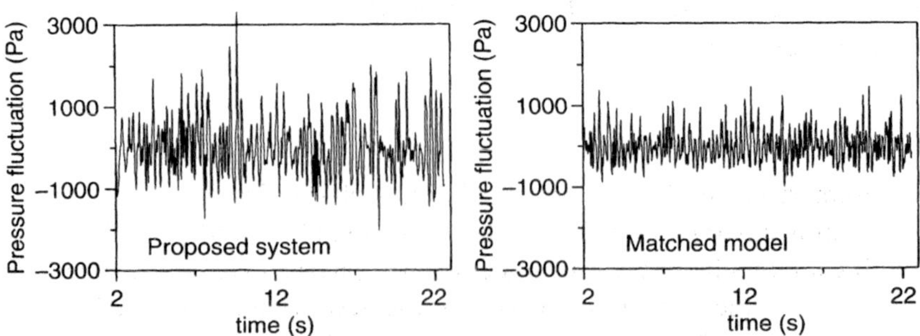

Figure 16.8 Fluid pressure fluctuations in the two matched systems.

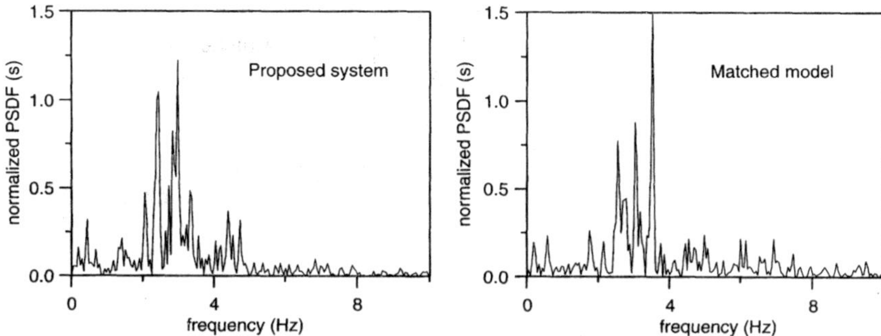

Figure 16.9 Power spectrum density functions (PSDF) for the two matched systems.

which reveals dominant frequencies within the 2–4 Hz range for both systems.

Bubble velocities

In addition to the pressure measurements, void fraction data were recorded at location B, and simultaneously at a position 1 cm directly above point B. Cross-correlation of these two responses enabled the bubble velocities between the two points to be measured. These turned out to be quite similar: 0.80 m/s for the proposed system, and 0.74 m/s for the matched ambient air system. Mean void fractions in this location were also very similar: 0.52 and 0.54 respectively.

Bubbling characteristics

Figure 16.10 shows the fully developed bubbling characteristics of the two matched units. They are very similar, with bubble dimensions in both cases growing to about half the bed width, giving rise to intense bed surface disruption.

Future developments

The numerical simulations presented in this chapter could be thought to represent a starting point for programmes of study into the prediction of fluidization quality in existing and envisaged physical systems. The potential for such methods is immense, as the few results presented above amply illustrate. The eventual goal is a numerical code, which could be applied with confidence to the elimination of much of the uncertainty at

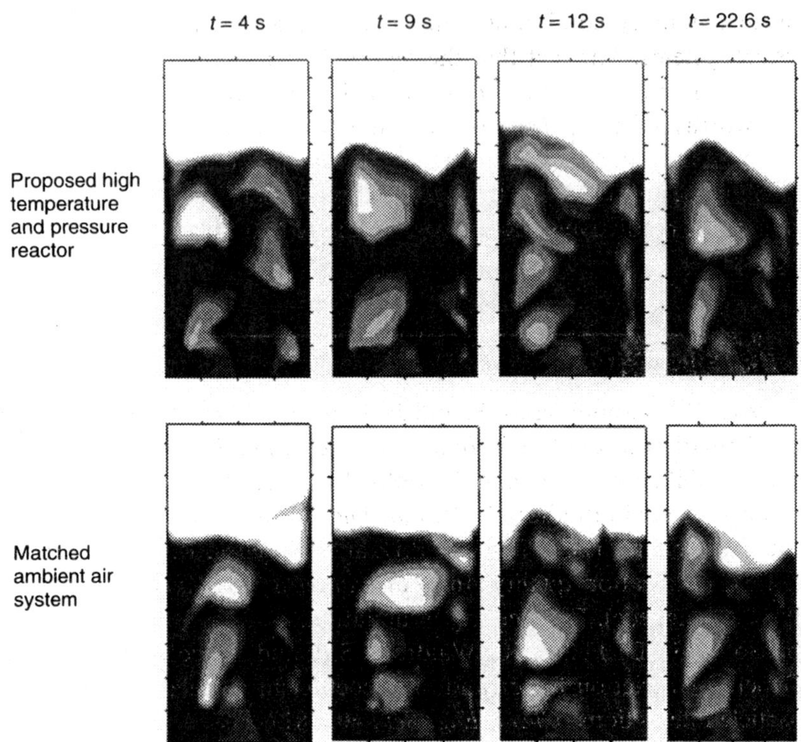

Figure 16.10 Fully developed bubbling characteristics of the two matched systems.

present inherent in the choice of a fluidization regime for operation under previously untested conditions.

Much of the emphasis of this book has been on the fully predictive nature of the constitutive relations adopted in the analysis. Arbitrary and adjustable parameters have been avoided completely, a point made from time to time along the way, but nevertheless worth repeating a final time here. The development is unique in this respect; and the fact that the one-dimensional formulation nevertheless homes in on good quantitative predictions of quite dramatic events – such as the spontaneous appearance of bubbles in a previously uniform particle suspension, accompanied by a sudden collapse in the bed height – testifies to its basic structural integrity.

The two-dimensional simulations presented in this chapter appear to capture well the essential features of bed behaviour, paving the way to a full three-dimension formulation, which would permit the introduction of realistic boundary conditions that reflect the geometric details of true physical systems. The fact that the reported achievements have involved only the unhindered particle drag-coefficient as empirical input is once again worthy of note. Moreover, although such extreme economy of empiricism is unlikely to continue once the precise details of experimental behaviour are confronted with matched three-dimensional simulations, at least a basic, predictive structure can be said to be in place from which to progress.

References

Anderson, K.S., Sundaresan, S. and Jackson, R. (1995). Instabilities and the formation of bubbles in fluidized beds. *J. Fluid Mech.*, **303**, 327.

Chen, Z., Gibilaro, L.G. and Foscolo, P.U. (1999). Two-dimensional voidage waves in fluidized beds. *Ind. Eng. Chem. Res.*, **38**, 610.

Chen, Z., Gibilaro, L.G., Foscolo, P.U. and Di Felice, R. (2001). Prediction of fluidization quality. In: *Fluidization X* (M. Kwauk, J. Li and W.-C. Yang, eds). Engineering Foundation.

Gibilaro, L.G., Di Felice, R., Waldram, S.P. and Foscolo, P.U. (1985). Generalised friction factor and drag coefficient correlations for fluid-particle interactions. *Chem. Eng. Sci.*, **40**, 1817.

Gidaspow, D., Syamlal, M. and Seo, Y. (1986). Hydrodynamics of fluidization of single and binary size particles: supercomputer modelling. In: *Fluidization V* (K. Ostergaard and A. Sorensen, eds). Engineering Foundation.

Glasser, B.J., Kevrekidis, I.G and Sundaresan, S. (1997). Fully developed wave solutions and bubble formation in fluidized beds. *J. Fluid Mech.*, **334**, 157.

Hassett, N.L. (1961). Flow patterns in particle beds. *Nature*, **189**, 997.

Author index

Subject index

Printed in the United Kingdom
by Lightning Source UK Ltd.
134274UK00003B/7-24/A

9 780750 650038